W0050885

IMPROVING GENETIC DISEASE RESISTANCE
IN FARM ANIMALS

Current Topics in Veterinary Medicine and Animal Science

Volume 52

For a list of titles in this series see final page of this volume.

Improving Genetic Disease Resistance in Farm Animals

A Seminar in the Community Programme for the
Coordination of Agricultural Research,
held in Brussels, Belgium, 8–9 November 1988

Sponsored by the Commission of the European Communities,
Directorate-General for Agriculture,
Coordination of Agricultural Research

Edited by

A. J. VAN DER ZIJPP

Department of Animal Husbandry,
Agricultural University,
Wageningen, The Netherlands

and

W. SYBESMA

Research Institute for Animal Production 'Schoonoord',
Zeist, The Netherlands

KLUWER ACADEMIC PUBLISHERS

DORDRECHT / BOSTON / LONDON

FOR THE COMMISSION OF THE EUROPEAN COMMUNITIES

ISBN-13: 978-94-010-6967-0   e-ISBN-13: 978-94-009-1057-7
DOI: 10.1007/978-94-009-1057-7

Publication arrangements by
Commission of the European Communities
Directorate-General Telecommunications, Information Industries and Innovation, Scientific and
Technical Communications Service, Luxembourg

EUR 11964
© 1989 ECSC, EEC, EAEC, Brussels and Luxembourg

Softcover reprint of the hardcover 1st edition 1989

LEGAL NOTICE
Neither the Commission of the European Communities nor any person acting on behalf of the
Commission is responsible for the use which might be made of the following information.

Published by Kluwer Academic Publishers,
P.O. Box 17, 3300 AA Dordrecht, The Netherlands.

Kluwer Academic Publishers incorporates the publishing programmes of
D. Reidel, Martinus Nijhoff, Dr W. Junk and MTP Press.

Sold and distributed in the U.S.A. and Canada
by Kluwer Academic Publishers,
101 Philip Drive, Norwell, MA 02061, U.S.A.

In all other countries, sold and distributed
by Kluwer Academic Publishers Group,
P.O. Box 322, 3300 AH Dordrecht, The Netherlands.

Printed on acid-free paper

All Rights Reserved
No part of the material protected by this copyright notice may be reproduced or
utilized in any form or by any means, electronic or mechanical,
including photocopying, recording, or by any information storage and
retrieval system, without written permission from the copyright owner.

TABLE OF CONTENTS

Session 4: MHC and disease associations
Chairpersons: M. Vaiman and S. Lazary

Session 5: Immune response markers and disease resistance
Chairperson: E. Andresen

Session 6: General discussion
Chairpersons: W. Sybesma and A.J. van der Zijpp

This publication contains the proceedings of a seminar held in Brussels on November 8-9, 1988. The title of the seminar was "Reducing the costs of disease by improving resistance through genetics". The seminar was held as an activity of the Community Programme for the Coordination of Agricultural Research, 1984-1988.

Costs of disease depend on losses caused by morbidity, mortality and production decreases and on the costs of preventive measures including vaccination and medication. Production losses often contribute a major portion to the total costs. To reduce costs of disease preventive measures like vaccination, preventive medication and hygienic procedures are applied. Genetic resistance is an attractive preventive measure because of its consistent nature in the next generations, because it precludes veterinary services and because there are no side-effects. Constraints are the long term investment, relatively slow progress per generation (in combination with production traits) and the considerable lack of knowledge about inheritance of resistance mechanisms in farm animals.

This seminar was organised to review the research on the Major Histocompatibility Complex disease associations as a model in the above mentioned context. Other genetic disease approaches, i.e. direct selection for disease resistance, or immune markers were reviewed for the same reasons. The European Community, Scandinavia and Switzerland have long supported genetic research on disease resistance. The European founders of the MHC research of farm animals all participated in the seminar. New developments in immunology and molecular biology and the future policies on vaccine development and application also made this meeting timely.

The seminar was divided into four main sessions:
1) MHC serology and immunology
2) MHC polymorphism by protein chemistry and DNA-techniques
3) MHC and disease associations
4) Immune response markers and disease resistance

Researchers of different disciplines, working with different farm animal species, exchanged knowledge and discussed their experiences with regard to immunological and molecular techniques, experimental populations, statistical evaluation, interpretations of results and needs for further research and international collaboration.

The programme was planned and organised by Dr.Ir. A.J. van der Zijpp, Department of Animal Husbandry, Agricultural University, Wageningen and by Dr. W. Sybesma, Research Institute for Animal Production "Schoonoord", Zeist, The Netherlands.

The Commission of the European Communities wishes to thank participants, speakers, chairpersons and those who took part in the discussions.

SESSION 1

GENERAL ASPECTS

Chairperson: Dr. A.J. van der Zijpp

INTRODUCTION

A.J. van der Zijpp

Department of Animal Husbandry, Agricultural University
P.O. Box 338, 6700 AH Wageningen, The Netherlands

Disease resistance of farm animals can be obtained by vaccination and genetic improvement. A third possibility is eradication. Eradication may be a satisfactory method, but it also has negative implications for genetic improvement for production and other traits besides cost. Vaccinations can be very successful, but for some diseases there are also drawbacks. Vaccines sometimes cause pathological responses, often associated with stress, followed by secondary infections. Sometimes disease outbreaks occur despite vaccination. Vaccines may temporarily reduce growth. Vaccines are not applicable everywhere in the world, especially where storage conditions are unfavourable. Finally, application of vaccines is labour intensive. Especially for those diseases where vaccines don't function optimally, genetic resistance may present an alternative.

Before animal vaccines became widely available, improvement of genetic resistance was part of any animal breeding programme. Natural selection for disease resistance probably was significant, because neither vaccination nor hygienic measures were common practice in the pre-war era. Now, with all in - all out systems, isolation, hygienic measures and vaccinations, natural selection in breeding populations is reduced to a minimum. Virtually all animals are available for selection for production traits. Little is known about phenotypic and genetic relationships between production and disease resistance traits. What will happen to disease resistance in the long run? What happens to disease resistance in less favourable conditions e.g. pathogenic environments?

With recent developments in immunological research and the introduction of recombinant DNA techniques, the possibilities for improvement of genetic resistance in farm animals have gained widespread research support. A major area of interest in all farm animal species became the Major Histocompatibility Complex. For mammals the MHC research originated in the late 60's and 70's. In chickens the MHC originally was detected as the B-bloodgroup in the late 40's. In fish the presence of an MHC has now also been indicated. Research efforts in farm animals have

3

continually been inspired by the fast developing knowledge and understanding of the MHC of mouse and man. The role of the human and mouse MHC in immune regulation and its association with infectious diseases have stimulated research for disease markers and for vaccine development in farm animal species.

Other research workers have approached disease resistance via immune response parameters, pathological indicators or effects of disease on production traits. In both approaches the value of the markers or parameters has to be proven with challenge testing or epidemiologically supported population data on disease.

Effective resistance to disease is most likely obtained by the combination of vaccination and genetic resistance. Gavora and Spencer (1983) have repeatedly shown the combined advantage of genetically resistant strains and vaccination for Marek's disease in the chicken. Future vaccine development will increasingly be based on knowledge of the interactions between MHC-coded gene products, T-cell receptors and processed antigen. This may imply that knowledge about diversity of immune mechanisms in populations becomes imperative for genetic resistance and efficient response to vaccines.

The costs of disease have been estimated to be 10-20% of the total value of production. Genetic improvement of disease resistance can contribute to a reduction of these costs. For successful genetic improvement the basic mechanisms of resistance have to be known and genetic effects have to be sizable: major gene effects or high heritabilities.

The objectives of the seminar therefore are:
1. To provide an overview of current research on genetic resistance to disease across farm animal species.
2. To compare diverse approaches e.g. MHC associations versus immune response and/or pathological parameters as markers for disease resistance.
3. To weigh different approaches to MHC-typing i.e. serology, protein chemistry of gene products and DNA polymorphism.
4. To indicate important areas for future research and gain insight in practical application of present knowledge.

REFERENCES

Gavora, J.S. and J.L. Spencer (1983). Breeding for immune responsiveness and disease resistance. Animal Blood Groups and Biochemical Genetics 14: 159-180.

Klein, J. (1986). Natural History of the Major Histocompatibility Complex. John Wiley and Sons, New York, pp. 1-774.

BIOLOGICAL SIGNIFICANCE OF THE MHC

R.R.P. de Vries

Department of Immunohaematology & Blood Bank
University Hospital, P.O. Box 9600
2300 RC Leiden, the Netherlands.

ABSTRACT

In this review the biological significance of the MHC is discussed. The author confines himself to class I and II genes and their products. Class I and II molecules present processed antigens to T cells. This function is operative during the development of a self-tolerant T cell repertoire and during an ongoing immune response. Both MHC class I and II genes are extremely polymorphic. This polymorphism results in inter-individual differences in immune reactivity. Therefore these genes are so-called Immune response (Ir-) genes. The resulting differences in immune reactivity are due to differential binding of processed antigen to the products of these Ir-genes. The MHC polymorphism has been conserved in evolution and there is evidence that selection by infectious disease has been involved in this process. An explanation for this is presented, which amounts to the idea that MHC polymorphism is a very pragmatic answer to the unpredictable challenges of infectious diseases. One of the consequences of this type of life-insurance is that individuals with certain MHC alleles have an increased susceptibility to certain immuno-pathological diseases. These recent developments in immunogenetics may be applied to the development of sub-unit vaccines, have implications for the prevention of T cell mediated immunopathological diseases and may result in genetic manipulation aimed at introducing resistance genes in susceptible animals.

BIOLOGICAL SIGNIFICANCE OF MHC CLASS I AND II MOLECULES

In this review I will confine myself to MHC class I and class II genes and molecules. I realize that between the class I and II genes several genes are located that may be quite relevant for the student of the MHC. These genes include those coding for complement factors, tumor necrosis factor (TNF), collagen-like genes and a heat shock protein. Whether these genes may be considered to be part of the MHC is an interesting question which I will not try to answer here.

The structure of a human MHC (HLA) class I molecule is known in con-siderable detail (Bjorkman et al.,1987). In all probability the structure of class II molecules is very similar (Brown et al.). The molecules stick out of the cell membrane and both seem to have at their top a unique "groove" which is formed by a bottom of beta-pleated sheets and walls of alpha-helices. Because at least the best-documented function of these molecules is to present antigen to T cells (Schwartz, 1985), it is now generally assumed that processed antigens (peptides) will bind to the MHC molecules in this groove and thus be presented to the T cell receptor (Parham, 1988). However, why is that so? Why could T cells not see free antigen like B cells? I think there are two reasons for that. The first is that MHC molecules and self peptides play a crucial role in the development of the T cell repertoire. A crucial role in two ways: T cells need to recognize self MHC molecules in order to survive in the thymus (so-called positive selection) and secondly the T cells that have too high avidity receptors for self MHC molecules are deleted (negative selection) in order to develop tolerance for self (Marrack and Kapler, 1988; Janeway, 1988). So HLA molecules play a crucial role during the development of the T cell repertoire. The second reason why I think T cells may need HLA molecules is to focus the immune response. This explanation is intuitive and lacks formal evidence. While B cells have a receptor for antigen (antibody) which can be secreted and binds antigens in solution, T cells have to recognize antigen presented by a MHC molecule on an antigen presenting cell, which also has to have the capacity to process that antigen (Allen, 1987). So in other words: this would enable the T cell system to focus on antigen in the lymphoid organs where the cells are with which they have to interact in order to eliminate the antigen. We should realise that the T cell system is older than the B cell system. Thus the antibodies, with their capacity to circulate and bind antigen everywhere where that might be necessary and their capacity of somatic mutation that T cell receptors do not have, have been superimposed on the cell-mediated immunity.

BIOLOGICAL SIGNIFICANCE OF MHC CLASS I AND II POLYMORPHISM

The MHC system is very polymorphic, in fact the most polymorphic genetic system we know. This polymorphism is located around the groove that is binding the peptide: both on the beta-pleated sheets at its bottom and on the alpha-helices aligning it (Bjorkman et al., 1987; Brown et al.). This suggests that it may affect antigen-binding and presentation to the T cell receptor. This is indeed the case and results in a phenomenon which we call immune response genes (Benacerraf and McDevitt, 1972; Schwartz, 1985; Allen et al., 1987; Buus et al., 1987). This phenomenon is illustrated in the left part of figure 1 where I have depicted antigen presenting cells (APC) of three HLA different individuals which have different HLA molecules with different grooves which bind different peptides. Thus, individuals with different MHC types may respond to different antigens.

The MHC class I and II polymorphism has been conserved in evolution. Some alleles are found in different species and therefore must have arisen many millions of years ago (McConnell et al., 1988; Figueroa et al., 1988; Lawlor et al., 1988). THis suggests that selection has been involved to maintain the polymorphisms (Howard, 1988). Evidence for this has recently been obtained by examining the pattern of nucleotide substitutions, which showed a significantly increased rate of non-synonymous (amino acid altering) substitutions in the regions coding for the "groove" and its surroundings (Hughes and Nei, 1988). The best candidates for the selective forces maintaining this Ir-gene polymorphism are of course infectious diseases, but there is not much evidence for that. We have studied an experiment of nature that provides evidence for this assumption (de Vries et al., 1979; de Vries et al., in press). In the middle of the last century a group of 367 Dutch farmers emigrated to Surinam was attacked by an epidemic of typhoid fever that killed half of them and a few years later 20% of the survivors was killed by a yellow fever epidemic. We could bleed the descendants of the 40% that survived these two epidemics and compare the gene frequencies of 26 polymorphisms with those of the Dutch population that stayed in the Netherlands. Whereas the gene frequencies of most polymorphisms were not very different, a few polymorphisms showed differences in frequencies that were unlikely to be due to drift, among which HLA (de Vries et al., 1979). At that time we could only type for HLA class I polymorphisms and because the data suggested that stronger differences might be observed for frequencies of HLA class II alleles we recently went back to do class II typing. The most relevant results were: an absence of HLA-DR2 which has a gene frequency of 0.17 among Dutch controls and significantly increased frequencies of DR4 and DRw13 (de Vries et al., in press). DR2 is therefore associated with mortality, whereas the presence of DR4 and DRw13 seems to have conferred survival advantage during at least the typhoid fever epidemic. The latter two alleles share a T cell defined epitope which we therefore think may be an immune response product conferring resistance to typhoid fever.

WHAT IS THE BIOLOGICAL ADVANTAGE OF MHC CLASS I AND II POLYMORPHISM?

To my knowledge there is neither much evidence nor a good explanation for heterozygote advantage acting at the level of the individual. Perhaps the diversity of the products of the different loci should also have to be taken into account here. There is also not much evidence that the maintenance of MHC class I and II polymorphisms acts at the level of the population, but at least there is a good explanation for that. We have seen that MHC molecules and the self-peptides bound to them play an important role in the development of the T cell repertoire, which has to be tolerant to the self-antigens of an individual. This self-tolerance results in "holes in the repertoire" and/or antigen-specific suppression, which will have consequences for the response to foreign antigens. MHC polymorphism will result in "polymorphism" of these "Achilles heels", which is obviously advantageous at least for survival of the species. Thus MHC polymorphism may lead to a spreading of the necessary risk of each individual for infectious diseases and may be considered a form of life-insurance: a very pragmatic answer to the unpredictable challenges of infectious diseases.

MHC AND DISEASE

One of the consequences of this type of life-insurance is that individuals carrying certain alleles have increased susceptibility to certain immunopathological diseases. I will very briefly illustrate this using data that we have generated during the last years from leprosy patients (de Vries et al. II., in press).

Figure 1 just summarizes these data and the conclusions that we have drawn. Different individuals with different MHC molecules present different M. leprae epitopes to different T cell subtypes. In the majority of cases so-called protective epitopes to a T cell subtype that confers immunity and thus protects against leprosy. Certain HLA alleles preferentially present M. leprae epitopes to T cells that also confer immunity, but as a side-effect confer immunopathology which is called tuberculoid leprosy. Finally people with still another HLA-allele can present epitopes that are seen by suppressor T cells which suppress the T cells that provide immunity and results in a type of immunopathology, which is mainly mediated by antibodies and is called lepromatous leprosy. What are the implications of all this? In the case of leprosy we have

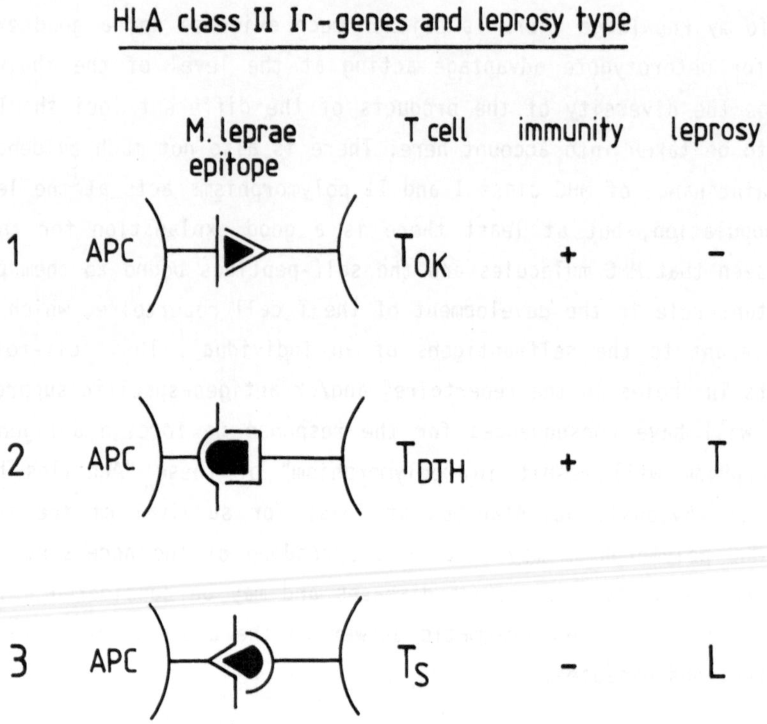

HLA class II Ir-genes and leprosy type

	M. leprae epitope	T cell	immunity	leprosy
1	APC ▷	T_{OK}	+	−
2	APC	T_{DTH}	+	T
3	APC	T_S	−	L

Figure 1: HLA class II Ir-genes and leprosy type. Antigen-presenting cells (APC) of three individuals differing for HLA class II present different M. leprae epitopes to functionally different T cells: T_{OK} is a helper T cell that confers protective immunity, the same is true for T_{DTH} which, however, also causes type IV immunopathology (delayed-type hypersensitivity (DTH)-reaction) seen in tuberculoid leprosy, and Ts is a suppressor T cell responsible for the M. leprae-specific non-responsiveness seen in lepromatous leprosy.

learned quite a lot about the pathogenesis of tuberculoid and lepromatous leprosy, which is basically immunopathology. In leprosy the challenge is of course to develop a vaccine which will protect from the disease. The ideal vaccine that we would like to develop would be a vaccine that does not contain the disease inducing epitopes, but only the protective (=immunity inducing) epitopes. But we have also to take into consideration that there will be also immune response gene control of the response to these protective epitopes. So in other words when we make a subunit vaccine, we have to make a vaccine that is seen by all individuals with all HLA types.

For T cell mediated immunopathological diseases in general several other approaches are possible, when we confine ourselves to the T cell receptor (TCR), MHC molecule and the disease-inducing peptide presented. If there is no relevant HLA molecule, the antigen will not be presented. So if we can down-regulate (preferentially locus-specific or allele-specific) MHC molecules which will present certain epitopes, we can prevent disease. A recent development is that it seems also possible to functionally block the groove of the MHC molecule which will present the antigen with a non-immunogenic peptide (Adorini et al., 1988). So injection of or otherwise providing the susceptible individual with a non-immunogenic peptide which will block the groove of the MHC molecule that can present a disease inducing epitope may prevent disease. If we know the disease-inducing peptide, we can try to induce tolerance with it. Of course we may also manipulate the T cell receptor and the T cell itself.

Finally, genetic manipulation may be applied to both MHC and T cell receptor genes if they are crucial in the resistance to certain diseases.

REFERENCES

Adorini, L. Muller, S., Cardinaux, F., Lehmann, P.V., Falcioni, F. and Nagy Z.A. 1988. In vivo competition between self peptides and foreign antigens in T-cell activation. Nature, 334, 623-625.

Allen, P.M. 1987. Antigen processing at the molecular level. Immunol. Today, 8, 270-273.

Allen, P.M., Babbit, B.P. and Unanue, E.R. 1987. T-cell recognition of lysozyme: the biochemical basis of presentation. Immunol. Rev., 98, 171-187.

Benacerraf, B. and McDevitt, H.O. 1972. Histocompatibility-linked immune resonse genes. A new class of egnes that controls the formation of specific immune response has been identified. Science, 175, 273-279.

Bjorkman, P.J., Saper, M.A., Samraoui, B., Bennett, W.S., Strominger, J.L. and Wiley, D.C. 1987. Structure of the human class I histocompatibility antigen, HLA-A2. Nature, 329, 506-512.

Bjorkman, P.J., Saper, M.A., Samraoui, B., Bennett, W.S., Strominger, J.L. and Wiley, D.C. 1987. The foreign antigen binding site and T cell recognition regions of class I histocompatibility antigens. Nature, 329, 512-518.

Brown, J.H., Jardetzky, T., Saper, M.A., Samraoui, B., Bjorkman, P.J. and Wiley, D.C. 1988. A hypothetical model of the foreign antigen binding site of class II histocompatibility molecules. Nature, 322, 845-850.

Buus, S., Sette, A., Colon, S.M., Miles, C. and Grey H.M. 1987. The relation between Major Histocompatibility Complex (MHC) restriction and the capacity of Ia to bind immunogenic peptides. Science, 235, 1353-1358.

De Vries, R.R.P., Meera Khan, P., Bernini, L.F., Van Loghem, E. and Van Rood, J.J. 1979. Genetic control of survival to epidemics? J.

Immunogenet., 6, 271-287.

De Vries, R.R.P., Schreuder, G.M.Th., Naipal, A, D'Amaro, J. and Van Rood, J.J. Selection by typhoid and yellow fever epidemics witnessed by the HLA-DR locus. Immunobiology of HLA vol. 2: "Immunogenetics and histocompatibility" (Ed. B. Dupont), Springer Verlag, New York, in press.

De Vries, R.R.P., Ottenhoff, T.H.M. and Van Schooten, W.C.A. HLA and mycobacterial disease. In "Immunology of Mycobacterial Disease" (Ed. P.J. Lachmann). (Springer Seminars in Immunopathology 10) in press.

Figueroa, F., Günther, E. and Klein, J. 1988. MHC polymorphism pre-dating speciation. Nature, 335, 265-267.

Howard, J.C. 1988. How old is a polymorphism? Nature, 332, 588-590.

Hughes, A.L. and Nei, M. 1988. Pattern of nucleotide substitution at major histocompatibility complex class I loci reveals overdominant selection. Nature, 335, 167-170.

Janeway, C.A. 1988. T-cell development. Accessories or coreceptors? Nature, 335, 208-210.

Marrack, P. and Kappler J. 1988. The T-cell repertoire for antigen and MHC. Immunol. Today, 9, 308-315.

McConnell, T.J., Talbot W.S., McIndoe, R.A. and Wakeland, E.K. 1988. The origin of MHC class II gene polymorphism within the genus Mus. Nature, 332, 651-654.

Lawlor, D.A., Ward, F.E., Ennis, P.D., Jackson, A.P. and Parham, P. 1988. HLA-A and B polymorphisms predate the divergence of humans and chimpanzees. Nature, 335, 268-271.

Parham, P. 1988. Presentation and processing of antigens in Paris. Immunol. Today, 9, 65-68.

Schwartz, R.H. 1985. Associations in T-cell activation. Nature, 317, 284-285.

SESSION 2

MHC serology and immunology

Chairperson: Dr. M. Simonsen

THE CHARACTERISATION AND FUNCTION OF THE BOVINE MHC

R.L. Spooner, R.A. Oliver, E.J. Glass and E.A. Innes

AFRC Institute of Animal Physiology and Genetics Research,
Edinburgh Research Station,
West Mains Road, Edinburgh, EH9 3JQ, Scotland, U.K.

ABSTRACT

The bovine MHC (BoLA) was first defined using serology. Sera from parous cows or from animals intentionally immunised or skin grafted were used in standard lymphocytotoxic assays. Some sera were absorbed with lymphocytes to render them operationally monospecific. Three international BoLA workshops have been organised and 34 internationally agreed specificities defined, all but two are clearly products of a single class I locus.

Alloreactive T cells recognise the same products as defined serologically and in the protozoan parasite diseases Tropical Theileriosis caused by Theileria annulata and East Coast fever caused by Theileria parva MHC restricted cytotoxic cells are generated and play an important protective role.

INTRODUCTION

The MHC in man and mouse comprises a series of closely linked genetic loci, which are the most polymorphic loci known. MHC gene products have been shown to play a vital role in directing and controlling the immune response, and can influence resistance and susceptibility to disease (Dausset and Svejgaard, 1978). A knowledge of the MHC will be necessary for the better understanding of pathogenesis, disease susceptibility and the development of vaccines particularly those comprising simple peptide antigens.

The characterisation of the MHC of domestic ruminants may constitute a first step towards increasing the efficiency of food production through improved disease resistance. Once identified, useful genes can be propogated rapidly in a population through the application of artificial insemination, embryo manipulation and DNA transgenic techniques. Improvements gained through such genetic approaches will reduce the need for sophisticated farm management, which is of importance in third world countries, where low input, low cost, robust systems often have a better chance of succeeding when intensive management and disease control systems are prone to collapse. Even where intensive livestock systems already exist, the potential for increased production through reductions in sub-clinical disease is too great to ignore.

CHARACTERISATION MHC CLASS I

Serology. The most widely used method for studying MHC class I
polymorphism is serology.   In man parous sera are used, while in cattle
alloantisera produced by skin grafting are favoured (Spooner, 1986).

The inheritance of serologically-defined bovine lymphocyte antigens
(BoLA) has been described (Spooner et al., 1978 and Amorena and Stone,
1978) and these have been shown to be controlled by alleles at a single
locus, the BoLA-A locus (Oliver et al., 1981).   A series of
international comparison tests have defined at least 34 BoLA-A locus
alleles (Bull et al., 1989) which compares with at least 50 alleles
identified serologically in mice (Table 1).   However the high frequency
of null alleles (undefined) in cattle shows that a considerable number of
bovine MHC class I gene products have yet to be identified.   As more
typing reagents become available the antigens presently characterised
will, in a number of cases, be 'split' resulting in the detection of
epitopes which are characteristic of individual gene products.   Such a
splitting of the A locus has been seen with the w6 specificity (Spooner
and Morgan, 1981: Anon, 1982) and the demonstration that w4, w7 and w10
form part of a cross-reactive group (Spooner, 1986).

Marked differences in BoLA gene frequencies are seen between European
breeds (Oliver et al., 1981), and also between breeds throughout the
world including the tropics (Spooner et al., 1987b; Kemp et al., 1988).
There is little serological evidence for more than one MHC class I locus
in cattle apart from two putative specificities identified in the 3rd
BoLA workshop (Bull et al., 1989) and data on ED74 (R.A. Oliver and R.L.
Spooner, unpublished).   This could be for two reasons: tight linkage of
the class I loci may result in sera identifying haplotypes, rather than
individual products.   There is some evidence for this in so far as three
workshop defined specificities are seen together in some breeds e.g. in
Hereford cattle w9, w13, w20 are often found together (R.A. Oliver and
R.L. Spooner, unpublished).   Also not all European sera work as well in
West Africa as in Europe for example only 6.4 behaves as a true w6
subgroup (Spooner et al., 1987b).   These examples may reflect
recombination between MHC class I loci which have been fixed in the
different breeds.   The sera produced in cattle may also identify only
the most antigenic locus.

TABLE 1.   BoLA specificities identified at the 3rd international BoLA
workshop

BoLA-A		BoLA-?	
New	Old/provisional	New	Old/provisional
w1	w1	w25[c]	a
w2	w2	w32[c]	.
w3	w3		
w4	w4		
w5	w5		
w6	w6		
w7	w7		
w8	w8		
w9	w9		
w10	w10		
w11	w11		
w12(w30)[b]	w12		
w13	w13		
w14(w8)[c]	w8.1		
w15(w8)[c]	w8.2		
w16	w16		
w17(w6)	w6.1		
w18(w6)	w6.1		
w19(w6)[c]	w6.4		
w20	w20		
w21[c]	.		
w22[c]	.		
w23(w5)[c]	w5.1		
w24[c]	.		
w26[c]	.		
w27(w10)[c]	w10.1		
w28[c]	.		
w29[c]	w28.1		
w30[c]	.		
w31[c]	w12.2		
w33[c]	.		

[a] symbol denotes no previous equivalent
[b] number in parenthesis is the supertype specificity
[c] new specificity defined by Third International BoLA Workshop

Monoclonal Definition of the Bovine MHC.   Although alloantisera have
been the basic tool for bovine class I antigen definition, monoclonal
antibodies have the advantage that they see individual epitopes and that
they can be produced in large quantities.   Several monoclonal antibodies
(mAbs) detecting epitopes on class I antigens of other species have been

tested with bovine cells, but most detect non-polymorphic determinants (Brodsky et al., 1981). Hitherto only a limited number of mAbs which define bovine class I polymorphism have been reported (Spooner and Pinder, 1983: Teale et al., 1986a; Chardon et al., 1983: Teale et al., 1986b; Teale and Kemp, 1987). It is nevertheless likely, that interest in the development of such reagents will be maintained. Evidence for other expressed MHC class I loci in cattle has been obtained from the sequential precipitation with several polymorphic bovine MHC class I monoclonal antibodies and the non-polymorphic human MHC class I monoclonal antibody, w6/32. Such an approach suggests that in excess of 4 different MHC class I antigens are found on the bovine cell surface (A. Bensaid and A.J. Teale, unpublished), the results are open to the criticism that each precipitation is incomplete.

T cell cytotoxicity. In the characterisation of the human MHC antigens alloreactive cytotoxic T lymphocytes (CTL) have been useful in detecting subtypes of serologically-defined class I antigen specificities (Horai et al., 1982: Spits et al., 1982). In cattle serologically defined subtypes also function as target antigens for alloreactive T cells (Teale et al., 1986a) whereas the supertypic specificities do not (Spooner et al., 1987a). There is better correlation between the CTL and serological definition of BoLA specificities than is found in man.

The derivation of alloreactive T lymphocyte clones specific for bovine class I antigens has recently been described (Teale et al., 1986b). While not useful as tools for routine typing, cloned T cells can be of use in fine dissection of selected haplotypes.

MHC restricted CTL are generated in infection with T. parva (Goddeeris et al., 1986) and with T. annulata and are an important protective mechanisms (Innes et al., 1989).

CLASS II ANTIGENS

Serology. Relatively little data is available on MHC class II serology in cattle. This is largely due to the lack of a suitable serological test. There are two principal reasons for this. First, class II antigens are only expressed on around 30% of a normal PBM population and second, most alloantisera possess anti-class I reactivity which must be removed before class II definition can be undertaken. This can be achieved by absorption with platelets and although the technique is tedious some useful sera have been reported (Mackie and Stear, 1988). Another method for separating cells involves the use of immunomagnetic granules. This has been used for class II typing in cattle but again is dependent on the production of suitable antisera (Lie et al., 1988).

A number of mAbs raised against MHC class II antigens of other species react with bovine class II products (Spooner and Ferrone, 1984; Lewin et al., 1985) and mAbs have been raised against the bovine antigens themselves (Letteson et al., 1983; Lalor et al., 1986) however, none has so far been shown to detect polymorphism.

T lymphocyte reactivity. In view of the difficulties involved in serotyping bovine class II antigens the use of cellular techniques has received greater emphasis. Using mixed lymphocyte reactions (MLR), in which T cells mount a proliferative response to non-self class II antigens, evidence was obtained suggesting significant polymorphism in bovine class II antigens (Usinger et al., 1977; Curie-Cohen et al., 1977). Further MLR studies, using full sib families demonstrated that the genes controlling the MLR were linked to the class I antigens (Spooner et al., 1978; Usinger et al., 1981).

As with serotyping methods for class II antigens, the MLR has not been widely used, due in part to the poor repeatability of results. recently the development of alloreactive bovine T cell clones has been reported (Teale et al., 1986b) and their use in the detection of class II antigens polymorphisms described (Teale and Kemp, 1987). Such T cell clones, characterised by the BoT4+ phenotype, mount reliable responses in proliferation assays. Their use is therefore not subject to the constraints affecting the standard MLR. They will probably be particularly appropriate for the definition of functionally important epitopes on MHC class II molecules, as in other species (Rosen-Bronson et al., 1986).

REFERENCES

Amorena, B. and Stone, W.H. (1978). Serologically defined (SD) locus in cattle. Science, 201, 159-160.

Anon. (1982). Proceedings of the Second International Bovine Lymphocyte (BoLA) Workshop. Animal Blood Groups and Biochemical Genetics, 3, 33-53.

Bull, R.W., Lewin, H.A., Wu, M.C., Pererbaugh, K., Antczak, D., Bernoco, D., Cwik, S., Dam, L., Davies, C., Dawkins, R.L., Dufty, J., Gerlach, J., Hines, H.C., Lazary, S., Leibold, W., Leveziel, H., Lie, O., Lindberg, F.G., Meggiolaro, D., Meyer, E., Oliver, R., Ross, M., Simon, M., Spooner, R.L., Stear, M., Teal, A., Templeton, J. (1989). Joint report of the Third International Bovine Lymphocyte Antigen (BoLA) Workshop. Animal Genetics. In press.

Brodsky, F.M., Stone, W.H. and Parham, P. (1981). Of cows and men: A comparative study of histocompatibility antigens. Human Immunology, 3, 143-152.

Chardon, P., Kali, J., Leveziel, H., Colombant, J. and Vaiman, M. (1983). Monoclonal antibodies to HLA recognise monomorphic and polymorphic epitopes on BoLA. Tissue Antigens, 22, 62-71.

Curie-Cohen, M., Usinger, W.R. and Stone, W.H. (1978). Transitivity of response in the mixed lymphocyte culture test. Tissue Antigens, 12, 170-178.

Dausset, J. and Svejgaard, A. (1978). HLA and Disease, (eds. J. Dausset and A. Svejgaard), Munksgaard, Copenhagen.

Eugie, E.M. and Emery, D.L. (1981). Genetically restricted cell-mediated cytotoxicity in cattle immune to Theileria parva. Nature, 290, 251-254.

Goddeeris, B.M., Morrison, W.I., Teale, A.J., Bensaid, A. and Baldwin, C.L. (1986). Bovine cytotoxic T cell clones specific for cells infected with the protozoan parasite Theileria parva: Parasite strain specificity and class I MHC restriction. Proceedings of the National Academy of Sciences (USA), 83, 5238-5242.

Horai, S., Poel, J.J. van der and Goulmy, E. (1982). Differential recognition of the serologically defined HLA-A2 antigen by allogeneic cytotoxic T-cells. I. Population studies. Immunogenetics, 16, 135-142.

Innes, E.A., Millar, P., Brown, C.G.D. and Spooner, R.L. (1989). The development and specificity of cytotoxic cells in cattle immunised with autologous or allogeneic Theileria annulata infected lymphoblastoid cell lines. Parasite Immunology. In press.

Kemp, S.J. Spooner, R.L. & Teale, A.J. (1988) A comparative study of major histocompatibility antigens in East Africa European cattle breeds. Animal Genetics 19, 17-30.

Lalor, P.A., Morrison, W.I., Goddeeris, B.M., Jack, R.M. and Black, S.J. (1986). Monoclonal antibodies identify phenotypically and functionally distinct cell types in the bovine lymphoid system. Veterinary Immunology and Immunopathology, 13, 121-140.

Letesson, J.J., Coppe, Ph., Lostrie-Trussart, N. and Depelchin, A. (1983). A bovine Ia-like antigen detected by a xenogeneic monoclonal antibody. Animal Blood Groups and Biochemical Genetics, 14, 239-250.

Lie, O., Vartdal, F., Funderud, S., Gaudernack, G., Froysadal, E., Olsaker, I., Ugelstad, J & Thorsby, E. Immunomagnetic isolation of cells for serological BoLA typing. Animal Genetics, 19, 75-86.

Mackie, J.T. and Stear, M.J. (1988). Serological definition of bovine major histocompatibility system class II gene products. Animal Genetics, 19, Suppl. 1, 10-12.

Morrison, W.I., Goddeeris, B.M., Teale, A.J., Baldwin, C.L., Bensaid, A. and Ellis, J. (1986b). Cell-mediated immune responses of cattle and their role in immunity to the protozoan parasite Theileria parva. Immunology Today, 7, 211-213.

Rosen-Bronson, S., Johnson, A.H., Hartzman, R.J. and Eckels, D.D. (1986). Human allospecific TLCs generated against HLA antigens associated with DR1 through DRw8. Immunogenetics, 23, 368-378.

Spits, H., Breuning, M.H., Ivanyi, P., Russo, C. and Vries, J.E. de. (1982). In vitro- isolated human cytotoxic T-lymphocyte clones detect variation in serologically defined HLA antigens. Immunogenetics, 16, 503-512.

Spooner, R.L., Leveziel, H., Grosclaude, F., Oliver, R.A. and Vaiman, M. (1978). Evidence for a possible major histocompatibility complex (BoLA) in cattle. Journal of Immunogenetics, 5, 335-346.

Spooner, R.L., Leveziel, H., Queval, R. and Hoste, C. (1987b). Studies on the histocompatibility complex of indigenous cattle in the Ivory Coast. Veterinary Immunology and Immunopathology, 15, 377-384.

Spooner, R.L. and Brown, C.G.D. (1980). Bovine lymphocyte antigens (BoLA) of bovine lymphocytes and derived lymphoblastoid lines transformed by Theileria parva and Theileria annulata. Parasite Immunology, 2, 163-174.

Spooner, R.L. and Morgan, A.L.G. (1981). Analysis of BoLA w6.
Evidence for multiple subgroups. Tissue Antigens, 17, 178-188.

Spooner, R.L. and Pinder, M. (1983). Monoclonal antibodies to bovine
MHC products. Veterinary Immunology and Immunopathology, 4,
453-458.

Spooner, R.L. and Ferrone, S. (1984). Cross reaction of monoclonal
antibodies to human MHC class I and class II products with
bovine lymphocyte subpopulations. Tissue Antigens, 24,
270-277.

Spooner, R.L. (1986). The bovine major histocompatibility complex. In
'The Ruminant Immune System in Health and Disease'. (ed. W.I.
Morrison), Cambridge University Press, Cambridge, 133-151.

Spooner, R.L., Innes, E.A., Millar, P., Webster, J. and Teale, A.J.
(1987a). Bovine alloreactive cytotoxic cells generated in
vitro detect BoLA w6 subgroups. Immunology, 61, 85-91.

Teale, A.J., Morrison, W.I., Goddeeris, B.M., Groocock, C.M., Stgg, D.A.
and Spooner, R.L. (1985). Bovine alloreactive cytotoxic
cells generated in vitro : target specificity in relation to
BoLA phenotype. Immunology, 55, 355-362.

Teale, A.J., Morrison, W.I., Spooner, R.L., Goddeeris, B.M., Groocock,
C.M. and Stagg, D.A. (1986a). Bovine alloreactive cytotoxic
T cells. In 'The Ruminant Immune System in Health and
Disease', (ed. W.I. Morrison), Cambridge University Press,
322-345.

Teale, A.J., Baldwin, C.L., Ellis, J.A., Newson, J., Goddeeris, B.M. and
Morrison, W.I. (1986b). Alloreactive bovine T Lymphocyte
clones: An analysis of function, phenotype and specificity.
Journal of Immunology, 136, 4392-4398.

Teale, A.J. and Kemp, S.J. (1987). A study of BoLA class II antigens
with BoT4+ T lymphocyte clones. Animal genetics, 18, 17-28.

Usinger, W.R., Curie-Cohen, H. and Stone, W.H. (1977).
Lymphocyte-defined loci in cattle. Science, 196, 1017-1018.

Usinger, W.R, Curie-Cohen, M., Benforado, K., Pringnitz, D., Rowe, R.,
Splitter, G.A. and Stone, W.H. (1981). The bovine major
histocompatibility complex (BoLA): Close linkage of the genes
controlling serologically-defined antigens and mixed lymphocyte
reactivity. Immunogenetics, 14, 423-428.

CURRENT STATUS OF SLA CLASS I AND II SEROLOGY

Birte Kristensen

Department of Animal Genetics

The Royal Veterinary & Agricultural University

1870 Frederiksberg C, Denmark

ABSTRACT

Eighteen SLA class I specificities have been internationally defined. In addition 13 specificities have been defined
nationally by highly correlated sera and in family studies.
Based on known recombinants and accumulated data from typing
of more than 700 informative families in several countries the
presently defined SLA specificities were tentatively allotted
to 3 class I allelic series; an internatinal nomenclature has
been established, accordingly.

In commmercial European breeds 41 SLA haplotypes have
repeatedly been demonstrated in families. When families of
national breeds were considered, additionally 62 haplotypes
could be demonstrated.

Particular haplotypes were found in all investigated
breeds, but in varying frequencies. In contrast other haplotypes were only found in some breeds and always in relatively
high frequencies. Commercial breeds seemed less polymorphic
than national breeds not subjected to intensive selection.

Class II SLA specificities have not been internationally
defined. Six specificites in French commercial breeds,
however, have been tentatively defined by serology.

INTRODUCTION

The first swine lymphocyte antigens (SLA), described, were
those coded for by SLA class I genes. Initially class I alloantisera were characterized by several independent groups
using different nomenclature (Vaiman et al., 1970, Hruban et
al., 1977 and Sachs et al., 1976).

Recently the first international comparison test was
organized with the purpose of confirming the previous results
and setting up an international nomenclature. The essential
results of this test are presented in the following.

The current knowledge of the distribution of SLA class I
haplotypes in commercial European and national breeds is outlined and possible influence of e.g. selection forces on the
frequencies of these is discussed. Finally, the present status
of some recent developments in class II serology will be
commented.

INTERNATIONAL DEFINITION OF SLA CLASS I SPECIFICITIES

Table 1 lists the laboratories contributing sera and their respective code names. SLA reagents were produced by alloimmunisation of the animals (Vaiman et al., 1970 and Kristensen et al., 1985). Lymphocytes from 264 unrelated Landrace or Large White pigs were chosen by each of the 4 laboratories.

TABLE 1. LIST OF LABORATORIES CONTRIBUTING SLA ANTISERA AND THE CELL PANEL

Abbreviation	Country	Town	No. of sera	No. of cells
CL	Czechoslovakia	Libechow	18	12
DB	Denmark	Berne*	35	0
DC	Denmark	Copenhagen	31	74
FJ	France	Jouy-en-Josas	61	117
FM	FRG	Munich	5	0
SB	Switzerland	Berne	4	62
US	United States	Iowa State University	3	0

*: Sera produced in Switzerland, but located in Copenhagen.

The results were analyzed by calculation of correlation coefficients and the x^2 tests for independence and allelism between each pair of reagents (Feingold, 1966). Serum clusters were constructed on the basis of correlation coefficients ranging from 0.70 to 1.00. As illustrated in Table 2, 18 specificities (W1 - 18) were internationally defined. Further technical details have been reported (Renard et al., 1988).

In addition to these 18 specificities, 10 additional specificities are recognized in current French breeds (FJ1,4,13 19,20,23,25 and 28), Danish breeds (DC31) and Swiss breeds (SB22). These specificities are recognized in several countries and 3 specificites (BM36,37 and 38) in Belgian breeds (Dr. Varewijck, personal communication).
The mentioned specificities are recognized in several countries and for most of them highly correlated sera are available, either locally or from at least 2 countries (Vaiman et al., 1988). Thus at present at least 30 SLA class I specificities can be recognized in the commercial breeds in Europe.

Table 2. Correlations between anti-SLA reagents and between SLA antigens and reagents in the First International Workshop

SLA	R	Reagents	++	+-	-+	n	r
W1	0.78	FJ1978	27	0	3	264	0.92
		FJ1060	27	0	7	264	0.86
		DB2936	27	0	15	263	0.76
		DC580*	26	1	16	263	0.76
W2	0.75	FJ1570	62	0	21	264	0.81
		FJ1103	62	0	26	263	0.77
		CL23	62	0	34	263	0.72
		DC268	62	0	31	200	0.71
W3	0.77	FJ1367	20	0	3	262	0.90
		CL164	20	0	9	257	0.79
		DC408	19	0	17	263	0.68
		FJ1609	20	0	22	264	0.64
W4	0.86	SB1746	7	0	0	257	0.93
		FJ742	7	0	1	264	0.86
W5	0.74	FJ4394	32	0	7	264	0.88
		FJ59	32	0	17	264	0.76
		SB1743	32	0	20	258	0.73

SLA	R	Reagents	++	+-	-+	n	r
W7	0.93	FJ1599	26	0	2	264	0.94
		DC523	26	0	3	263	0.92
		CL200	26	0	4	258	0.90
		DB5874*	25	1	6	262	0.85
		FJ997*	25	1	6	264	0.85
		DC532*	25	1	15	263	0.73
W8	0.77	DC124	51	0	1	263	0.98
		DC1	43	2	1	200	0.94
		FJ4073	51	3	21	264	0.79
		DC406*	43	3	16	263	0.79
		FJ771*	45	6	12	263	0.78
		CL214*	46	4	15	257	0.78
W9	0.75	DB6435	46	3	16	262	0.78
		FJ1203	45	0	24	253	0.75
		DB6567	39	6	14	252	0.74
		DB6436	48	1	28	262	0.72
		CL485*	49	0	36	257	0.68
		FJ861*	48	1	34	264	0.68
W10	0.71	DC67	24	0	3	262	0.92
		FJ1748	24	0	16	190	0.72
W11	0.71	FJ1620	72	0	5	263	0.94
		FJ1385	53	19	2	263	0.79
		DC138	72	0	46	263	0.67

SLA	R	Reagents	++	+-	-+	n	r
W13	0.81	DC121	11	0	0	263	0.95
		DB7117	11	0	5	185	0.78
		DB2311	10	1	3	261	0.78
		FJ4126	11	0	6	264	0.76
		DC5*	10	1	7	263	0.68
		FJ1739*	11	0	13	263	0.63
W14	0.70	DB6432	35	0	12	258	0.82
		CL431	35	0	21	258	0.74
		FJ860	35	0	37	264	0.63
		FJ1304*	29	6	22	264	0.62
W15	0.78	FJ868	36	0	6	263	0.90
		DC291	36	0	7	263	0.89
		DB2937*	28	3	8	201	0.79
		DC580*	31	5	11	263	0.75
W16	0.71	DB6570	19	0	6	263	0.84
		DB6141	19	0	9	252	0.78
		FJ1004	19	0	11	264	0.75
		DB2707*	19	0	24	263	0.61
W17	0.78	FJ884	16	0	4	202	0.85
		FJ4861	16	0	5	253	0.83
		DC57	16	0	5	201	0.83
		DC407*	15	1	8	263	0.74

Table 2. (continued) Correlations between anti-SLA reagents and between SLA antigens and reagents in the First International Workshop

SLA	R	Reagents	++	+-	-+	n	r
W6	0.73	FJ880	19	0	4	264	0.88
		FJA398	15	4	0	264	0.85
		DB11	19	0	8	263	0.80
		DB12	19	0	13	263	0.73
		DC412*	17	2	17	258	0.61

SLA	R	Reagents	++	+-	-+	n	r
W12	0.81	FJ1933	12	0	2	264	0.88
		CL393	12	0	3	258	0.85

SLA	R	Reagents	++	+-	-+	n	r
		FJ4347*	14	2	7	263	0.72
W18	0.80	FJ1532	12	0	2	264	0.88
		FJ4540	12	0	2	264	0.88
		SB5289	12	0	5	247	0.79

R: mean of the correlation coefficient between antisera chosen to define SLA antigens

r: correlation coefficient between antisera and SLA antigens

* antisera which were excluded to define SLA class I specificities

THE SLA HAPLOTYPE CHART AND ALLELIC ASSIGNMENT

Based on analysis of at least 700 back-cross families 62 different haplotypes have until now been recognized (Table 3). Of these 41 have been demonstrated in at least 3 different European countries either in the comparison test or reported in earlier studies (Kristensen et al., 1985). So far the remaining haplotypes have only been recognized in France, where considerable effort have been devoted to find new haplotype combinations in local French breeds such as Normand, Blanc de l'Ouest, Bayeux, Basque and Gascon (Renard et al., 1986).

TABLE 3. SEGREGATING SLA HAPLOTYPE CHART

Code no.	Loci A	Loci C	Loci B	Code no.	Loci A	Loci C	Loci B
*H1	W15	FJ1	B1	H32			B1
*H2	W10	W14		*H33		FJ1	B1
*H3	FJ20	W2	FJ23	*H34	W15	FJ1	FJ23
*H4	FJ13	W9		*H35	W8	W9	
*H5		W5	W4	H36		W5-W14	
*H6	FJ20	W5	FJ4	*H37	FJ27		
*H7	FJ20-W8	W2	B11	*H38	FJ33	FJ28	
*H8	W17			H39	W17	W16	
*H9		W5	FJ4	*H40	SB41		
*H10	FJ33	W12		H41			FJ23
*H11	FJ20-W7	W2		*H42	FJ29		
*H12	FJ13	FJ1	B3	*H43	W15	FJ1	
*H13	FJ13		B3	H44	FJ20-W7	W2	B11
*H14	FJ33	W16	B11	H45	W10		
*H15	FJ25			H46	FJ20-W7	W12	
*H16	FJ13	DB19		H47	FJ35		
*H17		DB19		*H48	W8		DC31
*H18		W5		H49	FJ20-W8	FJ1	
*H19			W6	H50		FJ1	
*H20	W13	W2	B11	*H51		W5-W14	B11
*H21	FJ20	W2	B11	H52		FJ1	B3
*H22	W8	W14		H53	FJ13	W9	W6
*H23	W18			H54	FJ20		
*H24			SB22	H55	FJ13	W9	B3
*H25	FJ20-W8	W2		H56		W5	W6
H26	FJ13	SB19	W6	H57	FJ20-W7		FJ23
*H27	W5	B11		H58	FJ20	W14	FJ23
*H28	FJ13		W6	H59	US line c		
*H29	FJ13			H60	FJ20-W8		
*H30		W9		*H61		W14	
H31	W8	W16	B11	*W62	FJ25	W14	

* Haplotypes recognized in at least 2 countries.

SLA HAPLOTYPE FREQUENCIES IN SELECTED EUROPEAN SWINE BREEDS

The frequencies of SLA haplotypes in selected European breeds are shown in Figs. 1 and 2. Some haplotypes, or combinations of genes are found in almost all the breeds i.e. H4 although with variable frequency. For example the H4 is a frequent haplotype in Swiss Landrace and in Danish Duroc.

Other haplotypes such as H7 seem to exist in a high frequency in all investigated Landrace breeds and in most breeds in very high frequencies (40%). Also Swiss and Danish Landrace seem less polymorphic than Belgian and French Landrace. The reason is at present speculative, but could be due to less differentiating class I typing in these countries, or to more intensive selection in these populations provided that selection influences the frequencies of haplotypes. The latter is supported by the observation that local French breeds not subjected to intensive selection displayed more polymorphism (Renard et al, 1986).

SLA CLASS II SEROLOGY AND CLASS II CHARACTERISATION BY MLC

SLA class II serology is still in its infancy. SLA recombinants (Vaiman et al., 1979) were used to raise anti-class II reagents. Preliminary cluster analyses enabled detection of at least 6 class II specificities (Vaiman et al., 1982), but the reagents were only moderately correlated (r values ranging from 0.4 to 0.6).

In Danish Landrace a segregation distortion of haplotype H7 has been demonstrated (Philipsen and Kristensen, 1985). Typing of back-cross families reveal that in the majority of sows or boars an excess of piglets inherit H7 (designated type H(+)). In approximately 10% of the H7 carrying parents, however, a decreased number of piglets inherit H7 (designated H7(-)).

As seen in Table 4 immunization of breeding animals with known H7 segregation pattern has resulted in production of antisera against the H7(+) type. Preliminary investigations have indicated that these sera recognize class II structures. (the reactivities are restricted to B-lymphocytes).

FIG 1: *SLA haplotype frequencies of selected european Landrace populations.*

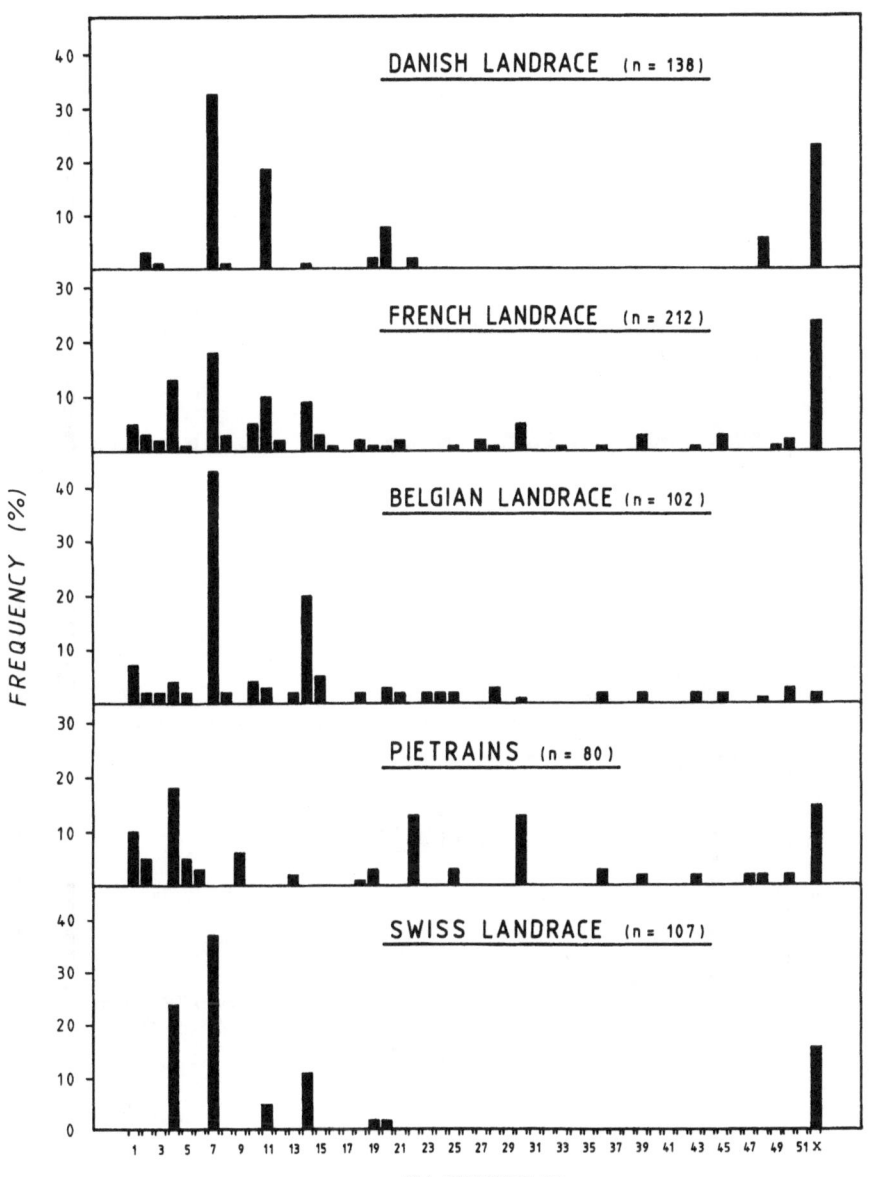

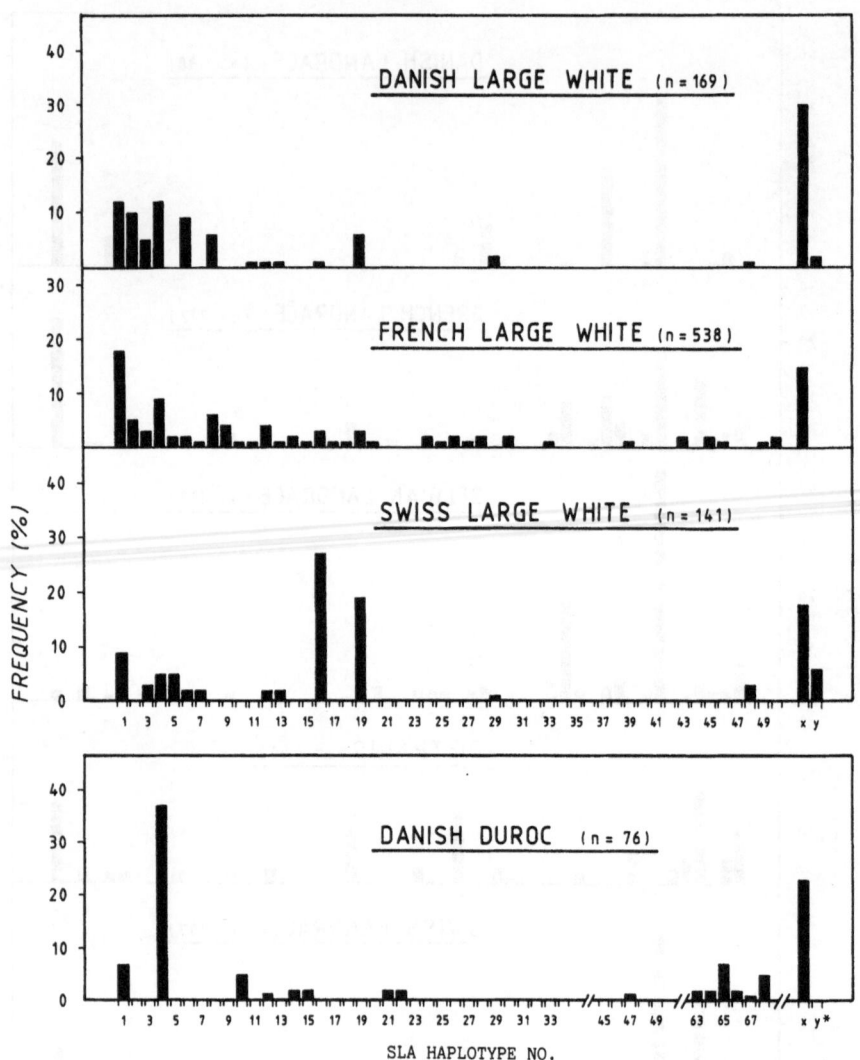

FIG 2: SLA haplotype frequencies of selected european Large white or Duroc populations.

*: Other haplotypes (frequency 1%).

subjected to intensive selection displayed more polymorphism (Renard et al, 1986).

SLA CLASS II SEROLOGY AND CLASS II CHARACTERISATION BY MLC

SLA class II serology is still in its infancy. SLA recombinants (Vaiman et al., 1979) were used to raise anti-class II reagents. Preliminary cluster analyses enabled detection of at least 6 class II specificities (Vaiman et al., 1982), but the reagents were only moderately correlated (r values ranging from 0.4 to 0.6).

In Danish Landrace a segregation distortion of haplotype H7 has been demonstrated (Philipsen and Kristensen, 1985). Typing of back-cross families reveal that in the majority of sows or boars an excess of piglets inherit H7 (designated type H(+)). In approximately 10% of the H7 carrying parents, however, a decreased number of piglets inherit H7 (designated H7(-)).

As seen in Table 4 immunization of breeding animals with known H7 segregation pattern has resulted in production of antisera against the H7(+) type. Preliminary investigations have indicated that these sera recognize class II structures. (the reactivities are restricted to B-lymphocytes).

TABEL 4. RESULTS OF IMMUNIZATION WITH SLA TYPE H7(+)

Donor		Recipient		
No.	SLA	No.	SLA	Antiserum
19	H7(+)/H7(-)	342	H7(-)/H2	Anti (+)
19	-/-	26	H7(-)/H48	Anti (+)

CONCLUDING REMARKS

As mentioned earlier, the common commercial breeds can to a large extent be SLA class I typed with the presently known panel of reagents. Eighteen SLA class I antigens identify at least 41 SLA haplotypes. Furthermore, it is probable that only a part of the SLA class I polymorphism has been discovered.

Recent development of powerful molecular methods have contributed to illucidate the genetics of the SLA. To gain optimal information, however, all available tools including classical serology should be used to further characterize the MHC and its products.

REFERENCES

Feingold, N. 1966. Utilisation du coefficiant de correlation dans l'etude des groupes leucocytaires. Rev. Franc. d'Etudes Clinique et Biolologiques, 11, 1036-1139.

Hruban, V., Simon, M., Hradecky, J. and Pazdera, J. 1977. Serologically defined specificities of the pig main histocompatibility complex (SLA). Anim. Blood Groups and Biochem. Genet., 8, 85-92.

Kristensen, B., Philipsen, M., Lazary, S., Renard, Ch. and de Weck, A.L. 1985. The swine histocompatibility system SLA: Serological studies in the common Swiss and Danish swine breeds. Anim. Blood Group and Biochem. Genet., 16, 109-124.

Philipsen, M. and Kristensen, B. 1985. Preliminary evidence of segregation distorsion in the SLA system. Anim. Blood Groups and Biochem. Genetics, 16, 125-133.

Renard, Ch., Luquet, M., Gaullieux, P. and Vaiman, M. 1986. Le polymorphisme du systeme majeur d'histocompatibilite SLA dans plusieurs races porcine en France. J. Rech. Porc. Fr., 18, 285-298.

Renard, Ch., Kristensen, B., Gautschi, C., Hruban, V., Fredholm, M. and Vaiman, M. 1988. Joint report of the First International Comparison Test on Swine Lymphocyte Alloantigens (SLA). Animal Genetics, 19, 62-72.

Sachs, D.H., Leight, G., Cone, J., Schwartz, S., Stuart, L. and Rosenberg, S. 1976. Transplantation in miniature swine. Transplantation, 22, 559-567.

Vaiman, M., Renard, Ch., LaFage, P., Amateau, J. and Nizza, P. 1970. Evidence for a histocompatibility System in Swine (SL-A). Transplantation, 10, 155-164.

Vaiman, M., Chardon, P. and Renard, Ch. 1979. Genetic organisation of the pig SLA complex. Studies on nine recombinants and biochemical and lysostrip analysis. Immunogenetics, 9, 353-361.

Vaiman, M., Renard, Ch., Chardon, P., Garin, B. and Leviziel, H. 1982. The D-DR region of the pig SLA complex. Analysis by serological and histogenetic methods. 18th Int. Conf. Anim. Blood Groups and Biochem. Polymorphism. Ottawa. pp. 15.

Vaiman, M. 1988. Histocompatibility systems in pigs. Prog. Vet. Microbiol. Immun., 4, 108-133.

ELA (EQUINE LYMPHOCYTE ALLOANTIGENS) SEROLOGY AND GENETICS

Varewyck H. and Y. Bouquet

State University of Gent
Department of Animal Breeding and Genetics
Heidestraat 19, B-9220 Merelbeke, Belgium.

ABSTRACT

As a result of 5 international workshops on Equine Lymphocyte Alloantigens (ELA) held between 1981 and 1987, 13 serologically defined specificities denoted as ELA-A1, A2, A3, A4, A5, A6, A7, A8, A9, A10, A14, A15 and A19 are attributed to locus ELA-A, and another 10 specificities denoted as ELA-W11, W12, W13, W16, W17, W18, W20, W21, W22 and W23 are known to segregate with ELA, but their position with regard to the ELA-A locus is yet unclair, except for ELA-W13, W22 and W23 which were assigned to a second locus called ELA-B. In the present paper, genetic evidence is presented for possible international acceptance of the specificities ELA-W16, W17, W18, and W20 and of 4 additional locally defined specificities, as ELA-A alleles. The distribution of all ELA-A antigens described here, was estimated in 5 Belgian horse populations.

Besides the ELA-antigens, other non MHC-specificities are internationally established as products of 2 independently segregating loci: ELY-1 and ELY-2.

SHORT HISTORY

The research on Equine Lymphocyte Alloantigens (ELA), which is the Major Histocompatibility Complex (MHC) of the horse, was started up in the mid-seventies by Swiss (Lazary et al., 1975; de Weck et al., 1978) and Anglo-American laboratories (Bright et al., 1978). From the very beginning, the study of the MHC in the horse was motivated by the finding of many MHC-disease associations in human. By now, associations are known between the presence of particular ELA specificities and an increased risk for diseases such as chronic bronchitis and laminitis (Lazary et al., 1982[a]), sweet itch (Lazary et al., 1982[c]) and sarcoid tumors (Dubath et al., 1986).

Between 1978 and 1980 several reports on lymphocyte alloantigens of the horse were published. Apart from the laboratories in Bern and Cambridge mentioned earlier, interest was taken in ELA research by different universities and institutions, most of them located in the United States. Generally, serologists involved in ELA research, exploited a most useful phenomenon in horse immunology: 80 up to 90 percent of the serum samples taken from primiparous mares shortly after parturation, do contain alloantibodies directed against ELA antigens. By means of appropriate dilution or absorption, such sera present the main source of reagents,

although immunizations with lymonocytes or whole blood have also been
performed successfully.

All labs made use of a two stage lymphocyte microcytotoxicity test
in order to define ELA class I specificities. But also the Mixed Lymphocyte Culture (MLC) has been applied in ELA, revealing the existence of a
class II regio (Lazary and Muller, 1980; Lazary et al., 1980). The ELA
regio was also shown to be linked to the bloodgroup A locus, which
created "linkage group III" (Bailey et al., 1979). Later, linkage was
also observed with the C4 and 21-hydroxylase genes (Kay et al., 1987).

THE INTERNATIONAL ELA WORKSHOPS

At the beginning of the eighties, many laboratories were working on
ELA, each of them having defined their own specificities using local
nomenclatures. In order to avoid a Babylonic confusion of tongues, a
first international workshop on lymphocyte alloantigens of the horse was
held in 1981 (Bull, 1983). Twelve different labs were participating
exchanging reagents for typing purposes. As a result, 6 serologically
defined ELA alloantigens were internationally accepted, denoted ELA-W1,
W2, W3, W4, W5 and W6. The conditions for defining new ELA alloantigens
were established as follows. First, serum cluster analysis on a lymphocyte panel, has to show an average correlation coefficient of at least
0.5. Second, each cluster should be composed by at least 3 different
reagents. Third, sera belonging to a same cluster, must originate from at
least two different laboratories. And of course, triplets may not be
observed. During the same workshop, the NIH-test (National Institute of
Health, Bethesda, Maryland) was accepted as the standard microcytotoxicity test to be applied in lymphocyte typing of horses.

The second international ELA workshop (Bailey et al., 1984) revealed
the existence of 4 more ELA specificities, denoted ELA-W7, W8, W9, and
W10, and of a non ELA lymphocyte antigen called ELY-2.1. This specificity
was found to be present on both T and B lymphocytes, but evenso on erythrocytes and thrombocytes (Antczak, 1984). The ELY-2:1 antigen segregates independently from the ELA locus, as well as from bloodgroup genes
and sex chromosomes (Bailey and Henney, 1984). Also during this workshop,
3 different methods for isolating lymphocytes from peripheral blood, were
compared. The methods in which thrombin or carbonyl iron were used in
order to deplete thrombocytes or granulocytes, were quite satisfactory,

but the method based on slow centrifugation was unable to produce pure
lymphocyte suspensions.

As a result of the third ELA workshop (Antczack et al., 1986), the
ELA-series was completed with specificity ELA-W11, whilst a second non
ELA lymphocyte antigen was also accepted, denoted ELY-1.1. Unlike
ELY-2.1, specificity ELY-1.1 is absent on thrombocytes and erythrocytes
(Lazary et al., 1982b) and was even exclusively detected on peripheral
T lymphocytes (Byrns et al., 1987).

In the fourth ELA workshop (Bernoco et al., 1987), blood samples of
different horse families were exchanged instead of reagents, which enabled to observe the segregation patterns of the ELA specificities. As a
consequence, the antigens ELA-W1 through W10 were shown to be segregating
at one particular locus called ELA-A, so that the nomenclature of these
specificities could be changed into the notation ELA-A1 through A10. On
the other hand, triplets were occuring with antigen ELA-W11, resulting in
the exclusion of specificity W11 from the ELA-A series. Not less than 10
additional ELA specificities were defined in this workshop: ELA-W12, W13,
W14, W15, W16,W17,W18,W19,W20 and W21. These antigens were kept "under
study", and more genetic research was required before they could be
classed or not classed under the ELA-A locus. Next to this, a second
allele at the ELY-1 locus was observed, called ELY-1.2.

TABLE 1 :Internationally recognized ELA and ELY specificities.

Alleles at the ELA-A locus	ELA specificities under study	Specificities at non-MHC loci
ELA-A1	ELA-W11	ELY-1.1
A2	W12	1.2
A3	W13*	
A4	W16	ELY-2.1
A5	W17	
A6	W18	
A7	W20	
A8	W21	
A9	W22*	
A10	W23*	
A14		
A15		
A19		

* assigned to locus ELA-B during the last ELA-Workshop (unpublished
    results)

At the fifth ELA workshop (report in preparation), two more ELA specificities were accepted: ELA-W22 and W23, and specificities W14, W15 and W19 were shown to be ELA-A alleles. This leads us to the present situation resumed in table 1. Thirteen ELA antigens are considered to be alleles at the ELA-A locus, and from the 10 ELA-W specificities under study, 3 antigens were classed in a second ELA-locus: ELA-W13, W22 and W23. The existense of this locus, called ELA-B, was proven by means of recombination data. The ELY-1 and ELY-2 alleles are segregating indepen-dently from each other and from linkage group III.

FURTHER SEROLOGIC AND GENETIC RESEARCH

At our laboratory in Merelbeke, Belgium, most of the internationally defined ELA and ELY antigens can be typed for by means of Belgian sera completed by sets of reagents from Switzerland (Prof. Dr. S. Lazary, Tierspital, Bern), France (Ir. G. Guérin, CNRZ, Jouy-en-Josas) and the United States (Prof. Dr. E. Bailey, University of Kentucky, Lexington, KY.). Apart from the internationally established antigens, four additio-nal ELA specificities were serologically and genetically defined as alleles segregating at the ELA-A locus: Be 7, Be 22, Be 25 and Be 108 (Varewyck et al., 1985). In table 2, a segregation analysis is shown, resulting from informative backcrosses observed in horse families belon-ging to different breeds being Belgian Warmblood, Belgian Trotter, Bel-gian Draft horse, Shetland pony, Arab and Thoroughbred. In this analysis all of the internationally established and of the four locally defined ELA specificities were followed, except for ELA-A15 which seemed to be poorly defined in our laboratory conditions, and for ELA-W22 and W23 for which no sera were available at that time. From table 2 it is clear that ELA-W16, W17, W18 and W20 can be considered as ELA-A alleles as well. This observation is confirmed by the population data listed in table 4, and by the fact that triplets were never observed. Further segregation analysis (results not shown) indicated that specificities ELA-W11, W12, W13 and W21 are inherited together with ELA-A alleles and do form triplets with the ELA-A series. It was indeed observed that ELA-W11 can segregate together with ELA-A14, A19, W20, Be 7 and the blank allele; ELA-W12 never segregates without ELA-A3; ELA-W13 segregates simultane-ously with ELA-A3 or A5; and ELA-W21 was observed to be inherited with ELA-A1, A2, A3, A4, A5, A7, A10, A14, Be 7 and with the blank allele.

At the Swiss lab (Hirni, 1988), the ELA-B specificity ELA-W13 was observed to segregate with ELA-A3 or A5; ELA-W22 segregates with ELA-A2, and ELA-W23 is inherited together with ELA-A5 or A15. Recombinations were observed between ELA-A alleles and all 3 of the ELA-B specificities as well as with the locally defined Swiss specificities Be VIII and Be 200 which might also be ELA-B alleles. Antigen ELA-W21 was provisionally assigned to a third locus called ELA-C, but no recombinations were observed. Specificities ELA-W13, W22 and W23 are absent on platelets and the homozygous combinations of W13 and W22 stimulate no heterozygous carriers. However, since these antigens are also present on T-lympho-cytes, it is not clair if ELA-B should be considered as a class II locus.

TABLE 2 : Backcrosses for ELA-A alleles in different horse populations.

	+	-		+	-
ELA-A1	34	30	W16	2	5
A2	84	97	W17	7	1
A3	73	75	W18	13	19
A4	10	13	W20	16	20
A5	25	15	Be 7	57	57
A6	20	12	Be 22	6	3
A7	13	14	Be 25	8	8
A8	9	17	Be 108	1	0
A9	14	7	Blank	82	84
A10	21	12			
A14	28	34			
A19	32	32			
			Total	555	555

$\Sigma \ X^2$ 7.79   not significant (q-1=16)

DISTRIBUTION OF ELA AND ELY ANTIGENS

Up to now, several papers on ELA and ELY gene frequencies in different horse breeds have been published. Table 3 shows a summary of ELA-A, ELY-1 and ELY-2 gene frequency estimations taken from the literature (references available on request). It is interesting to note that some specificities are rather typical for a given breed while others may be completely absent. E.g. specificities ELA-A1 and A10 have very high frequencies in American Standardbreds, ELA-A3 and A9 are abundantly represented in French Trotters, and ELA-A5 is often called "the Thorough-bred allele" because of its pronounced presence in this breed. On the

other hand, several antigens are absent in particular breeds, e.g. ELA-A1
in Thoroughbreds and Shetland ponies. As for the ELY alleles, ELY-1.1 is
quite frequent in all of the breeds investigated, but ELY-2.1 occurs at
much lower frequencies except for Shetland ponies.

The total ELA gene frequencies resulting from the data in table 3,
are most satisfactory for the American Standardbred (0.97), but are lower
in French Trotters (0.83) and Thoroughbreds (0.77). Extremely low values
were observed in the combined draft horses (0.52) , Iceland ponies (0.38)
and Shetland ponies (0.31). Therefore, at the Belgian laboratory five
different breeds were lymphocyte typed not only for the internationally
recognized ELA-A alleles, but also for the ELA-W specificities wich were
shown in this paper to segregate at the ELA-A locus, and for our locally
defined ELA-A specificities Be 7, Be 22, Be 25 and Be 108. The data
presented in table 4, confirm the low total gene frequencies observed in
draft horses and Shetland ponies when testing only for the established
ELA-A alleles. The total gene frequency in Belgian Warmbloods is also
low, but the figure found for Belgian Trotters is homologous to that
observed in French Trotters. On the other hand, the total gene frequency
observed by us in Thoroughbreds is higher than the values taken from the
literature. This is partly due to the higher frequence of ELA-A5 in the
Belgian Thoroughbred population. It is interesting to see what happens
when the ELA-W alleles segregating also at the ELA-A locus, are evenso
taken into account. In that case, the biggest progress is made in the
Belgian Warmblood (+0.077) thanks to the presence of ELA-A18 and A16. No
considerable progress is made in the other breeds. When our locally
defined ELA-A specificities are also considered, the total gene frequencies score always higher than 90% except in the Belgian Draft horse. Of
particular importance are specificity Be 25 in Shetland ponies and Be 7
in Belgian Draft horses, Belgian Warmbloods, Belgian Trotters and
Thoroughbreds. It should also be noticed that the ELY-2.1 antigen is
almost absent in the Belgian Draft horse.

This makes the ELA-A system one of the most powerful tools in paternity testing. Combined probabilities of exclusion (Bailey, 1984) were
calculated in each of the five breeds presented here. The CPE values
ranged between 0.67 in Shetlands and Belgian Draft horses up to 0.82 in
Belgian Warmbloods.

TABLE 3   Allele frequencies for loci ELA-A, ELY-1 and ELY-2,
in Thoroughbred. Standardbred and coldblood horses.

population	A1	A2	A3	A4	A5	A6	A7	A8	A9	A10	ELY-1.1	ELY-2.1
American Thoroughbred	.00	.15	.18	.00	.20	.02	.01	.00	.16	.05	.30	.06
American Standardbred	.26	.02	.04	.10	.06	.10	.06	.07	.01	.25	.32	.07
American Trotter	.34	.00	.05	.09	.04	.17	.01	.01	.00	.27	NR	NR
American Pacer	.23	.00	.03	.09	.12	.04	.05	.06	.00	.34	NR	NR
French Trotter	.05	.01	.30	.01	.04	.01	.01	.05	.30	.05	.17	.10
Combined draft horses	.05	.15	.16	.08	.00	.01	.00	.06	.00	.01	.11	.06
Shetland pony	.00	.03	.11	.01	.00	.14	.01	.00	.00	.01	.27	.32
Iceland pony	.05	.11	.08	.01	.02	.02	.00	.09	.00	.00	.41	.09

TABLE 4   Gene frequencies for ELA-A alleles and non ELA-A
specificities, and for the ELY-alleles, estimated in five
Belgian horse populations.

	Belgian Warmblood	Belgian Trotter	Belgian Drafthorse	Shetland pony	Thorough-bred
	n=394	n=223	n=296	n=100	n=364
ELA-A1	0.023	0.183	0.024	0	0
2	0.185	0.009	0.203	0	0.171
3	0.175	0.194	0.122	0.094	0.209
4	0.055	0.037	0.043	0	0
5	0.072	0.037	0	0	0.266
6	0.049	0.032	0	0.252	0.010
7	0.011	0.060	0	0	0.004
8	0.013	0.023	0.061	0	0
9	0.034	0.140	0	0.036	0.137
10	0.013	0.112	0	0	0.052
14	0.011	0	0.120	0.062	0.003
19	0.088	0.018	0.002	0.169	0.042
subtotal	0.729	0.845	0.575	0.613	0.849
W16	0.022	N.T.	0	N.T.	0
17	0	0	0.027	0	0
18	0.041	0.002	0	0.010	0
20	0.014	0.005	0	0	0.022
subtotal	0.806	0.852	0.602	0.623	0.916
Be 7	0.089	0.053	0.144	0.020	0.057
22	0.017	0	0.003	0.036	0
25	0.001	0.023	0.024	0.272	0
108	0	0	0.009	N.T.	0
Total	0.913	0.928	0.782	0.951	0.973
ELY-1.1	0.309	0.254	0.115	0.337	0.503
ELY-2.1	0.088	0.092	0.002	0.360	0.052

REFERENCES

Antczak, D.F. 1984. Lymphocyte alloantigens of the horse. III. ELY-2.1:
     a lymphocyte alloantigen not coded for by the MHC. Anim. Blood Grps
     biochem. Genet. ,15, 103-115.
Antczak, D.F., Bailey, E., Barger, B., Bernoco, D., Bull, R.W., Guérin,
     G., Lazary S., Mc Clure, J., Mottironi, V.D., Symons, R., Templeton,
     J. and Varewyck, H., 1986. Joint report of the 3rd international
     workshop on lymphocyte alloantigens of the horse, held 25-27 April
     1984. Anim. Genet., 17, 363-373.
Bailey, E. 1984. Usefulness of lymphocyte typing to exclude incorrectly
     assigned paternity in horses. Am. J. Vet. Res., 45, 1979-1983.
Bailey, E., Antczak, D.F., Bernoco, D., Bull, R.W., Fister, R., Guérin,
     G., Lazary, S., Matthews, S., Mc Clure, J., Meyer, J., Mottironi,
     V.D. and Templeton, J. 1984. Joint report of the 2nd international
     workshop on lymphocyte alloantigens of the horse, held 3-8 October
     1982. Anim. blood Grps biochem. Genet., 15, 123-132.
Bailey, E.and Henney, P.J. 1984. Comparison of ELY-2.1 with blood group
     and ELY-1 markers in the horse. Anim. Blood Grps biochem. Genet.,
     15, 117-122.
Bailey, E., Stormont, C., Suzuki, Y. and Trommershausen-Smith, A. 1979.
     Linkage of loci controlling alloantigens on red blood cells and
     lymphocytes in the horse. Science, 204, 1317-1319.
Bernoco, D.,Antczak, D.F., Bailey, E., Bell, K., Bull,R.W., Byrns, G.,
     Guérin, G., Lazary, S., Mc Clure, J. and Varewyck, H. 1987. Joint
     report of the 4th international workshop on lymphocyte alloantigens
     of the horse, held 19-22 October 1985. Anim. Genet., 18, 81-94.
Bright, S., Antczak, D.F. and Ricketts, S. 1978. Studies on equine
     lymphocyte antigens. Proc. 4th Confer. Equine Infect. Diseases, 229-
     236.
Bull, R.W. (Ed) 1983. Joint report of the 1st international workshop on
     lymphocyte alloantigens of the horse, held 24-29 October 1981.
     Anim. Blood Grps biochem. Genet., 14, 119-137.
Byrns, G., Crump, A.L., Lalonde, G., Bernoco, D. and Antczak, D.F., 1987.
     The ELY-1 locus controls a di-allelic alloantigenic system on equine
     lymphocytes. J. Immunogenet., 14, 59-71.
de Weck,A.L., Lazary, S., Bullen, S. Gerber, H. and Meister, U. 1978.
     Determination of leucocyte histocompatibility antigens in horses by
     serological techniques. Proc. 4th Confer. Equine Infect. Diseases,
     221-227.
Dubath, M.L., Gerber, H., and Lazary, S. 1986. Association between sus-
     ceptibility to equine sarcoid and ELA haplotypes in multiple-case
     families. Proc. 20th ISABR Confer., Helsinki, p. 28.
Hirni, H., 1988. The major histocompatibility complex in horses: ELA
     locus B and C antigens studied in unrelated animals and in families.
     DVM dissertation, University of Bern, 45 pp.
Kay, P.H., Dawkins, R.L., Bernoco, D., Tabarias,H. and Christiansen, F.T.
     1987. The genes for the fourth component of complements and the
     enzymes 21-hydroxylase are linked to the MHC in the horse.
     Anim. Genet., 18, suppl. 1, 78-79.
Lazary, S., Bullen, S., Muller, J., Kovacs, G., Bodo, I., Hockenjos, P.
     and De Weck, A.L. 1980. Equine leucocyte antigen system. II. Sero-
     logical and mixed lymphocyte reactivity studies in families.
     Transpl.,30, 210-215.
Lazary, S., de Weck,A.L., Straub, R. and Gerber, H. 1975. Characteri-
     zation of lymphocyte alloantigens permitting identification of horse

families. Proc. 1st Intern. Symp. Equine Hematol. (Michigan State University), p 132.

Lazary, S., Gerber, H. and Arnold, P., 1982^a. Distribution of leucocyte antigens in various pathological conditions in horses. Proc. 18th ISABR Confer., Ottawa, p. 66.

Lazary, S., Gerber, H., de Weck, A.L. and Arnold, P. 1982^b. Equine leucocyte antigen system. 3. Non-MHC linked alloantigenics system in horses. J. Immunogenet., 9, 327-334.

Lazary, S;, Larsen, H.J., Glatt, A., Isenbugel, E. and Gerber, H. 1982^c. Distribution of ELA antigens in "sweet itch" of horses. Proc. 18th ISABR Confer., Ottawa, p. 67.

Lazary, S. and Muller, J. 1980. Equine leucocyte antigen (ELA) system. II. Mixed lymphocyte reactivity studies. Proc. 17th ISABR Confer., Wageningen, p. 24.

Varewyck, H., Bouquet, Y., Lazary, S., Guérin, G., Van de Weghe, A. and Van Zeveren, A. 1985. Equine lymphocyte antigens in four major Belgian horse populations. Contribution to serology and antigen distribution. Anim. Blood Grps biochem. Genet., 16, 217-228.

THE CHICKEN MHC AND ITS IMPORTANCE[+]

M. Simonsen, M. Arnul*, P. Sørensen*
Institute for Experimental Immunology
University of Copenhagen
71 Nørre Allé
DK-2100 Copenhagen Ø, Denmark
*National Institute of Animal Science
Foulum, Post Box 39, DK-8830 Tjele, Denmark

ABSTRACT

MHC (B-complex) typing in chickens is usually done by haemagglutination which has the undisputed advantage of speed and ease, but also carries pitfalls that can lead to great confusion. Standard test sera in general use are nearly all defining B-haplotypes of White Leghorn and are not adequate for testing of other races, as exemplified by their use in heavy races used in the production of broilers. A development of alloantisera for typing of White plymouth Rock and White Cornish is described. A brief review is given of recent work relating genetic disease resistance in chickens to its B-complex.

INTRODUCTION

Chickens come in a great variety of shape, size, and color, and most of this variation is a product of civilization: the result of selection for desired characters including funny traits to suit the fancier's taste. By authority of Darwin (cited by Hutt, 1949), the origin of all domestic chicken is the wild Bankiwa chicken, or Red Jungle Fowl, of South-East Asia, although this monophyletic origin is not universally accepted.

The modern era of the chicken industry has seen two major selective pressures applied with unquestioned skill and a competitive zest bordering on the ruthless: one leading to excessive production of eggs in conditions of minimal space, another to excessively fast growth in meat-type chickens. Less deliberately, some amount of selection may long have occurred in both egg-layers and broilers against morbidity that interfered with the desired performance. This is now the type

+Composite paper covering 2 presentations at the meeting by Morten Simonsen, who is also accountable for the opinions expressed.

of selection which is deliberately coming to the fore: indeed, it is the theme of this meeting. Selection for good taste remains less conspicuous.

All these quantitative, commercially desirable traits are multigenic to the extent that they are inheritable, and it has long been a vexed question whether qualitative genetic markers such as blood groups were of practical use to the breeder.

It must be well-known history in this audience that the "B-complex" began as a "B-locus", i.e. one among several blood group systems, long before it emerged as the MHC of the chicken. However, once attained, the MHC status lends automatically support to the credibility of the B-complex being at least one of the sites of genetic polymorphism of importance for disease resistance. Obviously, analogies are being drawn to HLA in man, for which a great number of disease associations with different MHC alleles have been firmly established, although the exact mechanisms causing these associations are not yet understood.

In the following I will briefly discuss some of the better established associations between B-complex haplotypes and resistance to poultry diseases. My main emphasis will, however, be on the serology of the B-complex, which forms the basis of disease resistance studies as carried out so far. In so doing, I will put much emphasis on pitfalls and shortcomings that are common in B-complex serology, and offer suggestions for improvements.

THE B-COMPLEX STRUCTURE AND TERMINOLOGY

With the arrival of recombinant DNA technology to the small chromosome harbouring the chicken MHC, there is now rapid progress in delineating the structural genes of the B-complex: their numbers according to MHC class, their relative positions, etc. Here, Charles Auffray will bring us up to date at this meeting. Meanwhile, the business of B-complex serology goes on as usual. It remains essentially based on haemagglutination, and for discussion of the practical problems posed today in this field, the old 3-locus model of Pink et al.

(1977) is still useful.

Let me remind you that the 3 loci of this model belong to 3 different classes of MHC: class I (B-F), class II (B-L), andclass IV (B-G), whereas no class III (complement genes) have yet been identified. For general review, see Crone and Simonsen (1988).

In terms of both cellular expression and biochemical structure, both class I and class II products of the B-complex are clearly similar to the mammalian homologues. They differ from these in being discouragingly difficult to separate by crossing over (Skjødt et al. 1985), a fact which is actually getting a very rational explanation from the genomic organization now emerging (Auffray, this volume).

Recombination does occur, however, by crossing over between class IV and the 2 other loci, as first demonstrated by Hala et al. (1976). Although the frequency with which such recombinant chromosomes arise is no higher than about 1:2000 (Koch et al. 1983), they have been very valuable in serology, and still are, because they facilitate the production and control of antisera against one or the other of the 2 regions separated by the crossing over. This is particularly important since both class I and class IV are expressed in erythrocytes and are about equally highly polymorphic (Simonsen et al. 1982). As we shall soon see, the ability to discriminate by blood group serology between the antigens localized to class I (B-F) and those carried by class IV (B-G) is essential to avoid shear confusion.

By international agreement (Briles et al. 1982), the B-haplotypes are designated numerically, B1-n, and each comprises an allelic form of B-F (class I), B-L (class II) and B-G (class IV), denominated as the haplotype. Abbreviating these genes (or clusters of genes) to F, L, and G respectively, a B-haplotype has thus the general formula of
Bn = Fn, Ln, Gn. Since the L gene(s) is not expressed in erythrocytes, the haplotypes determinable by haemagglutination is really only the Fn, Gn part of the haplotype.

PITFALLS IN B-COMPLEX SEROLOGY

As mentioned above, recombinants separating F and L are so rare as to be negligible in a closed population. For this reason it is tempting to assume that all individuals in a closed population that carry the same F allele will also be identical in respect of L. Unfortunately, this may often be untrue, as RFLP-typing with L probes is already beginning to show. Presumably the most common cause for such discrepancies will prove to be that there are just many more haplotypes than we had thought. Some of them may differ only by minor changes in one of the genes. If this happens to be an L gene, the difference will clearly not be detectable by haemagglutination, and may even not be seen with antibodies to lymphocytes (immunofluorescence). Furthermore, minor changes may not be of minor importance, as illustrated by the fact that the great majority of human sufferers from juvenile diabetes carry a point mutation changing but one amino acid in the DQ β chain, which is a human class II product (Todd et al. 1987).

If typing for F cannot be relied upon as an indirect way of typing also for L, what then about typing for G alone? This is a practice which has been advocated and is often followed, probably because monoclonal antibodies for some of the classical G-alleles of White Leghorn are commercially available. The practice has been justified, though not by us, on grounds of the extremely strong linkage disequilibrium between F and G found by Simonsen et al. (1980) in outbred birds in respect of 3 of the standard haplotypes (B6, B12, and B13). I see no reason to assume a priori that typing for G should be better correlated with L than is typing for F.
On the contrary, one would expect, since crossing-over with G to one side and F and L to the other does after all occur with a measurable frequency (1:2000).

The shortcomings of B-typing by haemagglutination in White Leghorn are trivial, however, compared to the total confusion that may result from trying to apply the standard haemagglutinating test sera outside their own "province".

What then is their "province"? For historical reasons, it is the breed of White Leghorn, the famous and dominating egg-layers of the industry which may all originate (short of illigitimate later admixtures of genes) from one shipload of chickens that sailed from the Italian part of Livorno to North America in 1835. As to be expected, the MHC polymorphism of the species is vastly in excess of such a sample. This fact would not be so dangerous, were not haemagglutinating typing sera also cross-reactive. Unfortunately, that they are, sometimes in the extreme (e.g., an anti G12 serum raised in a strain combination precluding contamination with non-B antibodies will nevertheless agglutinate, more or less strongly, blood of all other haplotypes than that of the antibody producer).

In practice, one can usually reduce the nuisance of cross reacting antisera within their "province" to a tolerable degree by judicious choice of donor/producer combinations, combined with absorptions with lightly, or moderately cross-reacting haplotypes. In the end, a panel of reasonably monomorphic F and G reagents, or at least with only a few known, specific cross-reactions can be obtained that allow a quite precise determination of F,G haplotypes inside the "province". The moment one takes the same panel of test sera to a new "province", one can get into dire trouble. This fact will be illustrated in the following, moving from egg-layers to broilers.

B-HAPLOTYPES IN DANISH BROILERS

Danish broilers are Danish in that they live and die in Denmark, but their genes stem mostly from recent imports (1960ies and 1970ies) of American material (Sørensen, 1986) belonging to White Plymouth Rock (P.R.) and White Cornish (C.). Both of these races were developed in the U.S. in the late century. P.R. is of extremely mixed origin, while C. owes a great deal of its genome to Indian fighting cocks.

Our first attempts to type Danish broilers, both P.R. and C., by use of anti-F and anti-G sera defining international standard haplotypes soon convinced us of the futility of the exercise, but also provided the necessary background for selecting cocks to sire offspring from which new haplotypes could be isolated. Table 1 shows the typing data for the 11 cocks (6 P.R. and 5 C) that were selected to sire new MHC-families. In the right hand column is listed the final B-complex genotype of the cocks. It is seen that only one standard haplotype was found (B13 in cock C 3380). The problem is how to get from the main part of the table to the right hand column, and this will now be described in very broad outline. The detailed report by Arnull and Simonsen earlier announced (Simonsen 1987) is unfortunately still in pre-paration.

The 11 cocks of Table 1 were chosen on basis of their cross-reaction pattern with standard sera, supposing that most, or all would be heterozygous, and aiming of course at a broad coverage of the haplotypes segregating in the 2 populations.

The 11 cocks were then used to sire each 3 hens of White Leghorn, homozygous for either B15, or B19. (A 12th cock of unique cross-reaction pattern proved unfortunately sterile). Offspring of 20-30 chickens of each "cock family" were reared and tested with selected cross-reactive F and G antisera non-reactive with the maternal haplotype. All of the "cock families" segregated in ratios compatible with 1:1, thus supporting the expected B haplotype heterozygocity of the fathers. Not unexpectedly, some of the putative new haplotypes seemed to be repeats, which fact permitted reduction of the "cock families" to 7 (812, 818, 850, 1255, 3261, 3380, and 3612).

The 14 haplotypes segregating in these 7 families also comprised some repeats as indicated by serology. However, they might differ by other criteria, in particular such that are primarily, or wholly indicative of class II differences. Both direct compatibility measured by the graft-versus-host (GvH) splenomegaly reaction in chicken embryos, and the mixed lymphocyte reaction (MLR) constitute in our experience such

TABLE 1 Agglutinations of 11 broiler cocks with standard test sera

| Cock number | Specificity of standard test sera | | | | | | | | | | Final B genotype |
	F6	G6	F9	G12	F13	G13	F19	G19	F21	G21	
RR 812	++	++	-	-	-	-	-	-	-	-	127/129
PR 818	++	-	-	++	-	-	-	-	-	-	127/123
PR 850	-	-	(+)	++	-	-	(+)	-	-	-	125/123
PR 1203	-	++	(+)	++	-	-	-	-	-	-	125/129
PR 1242	-	++	++	++	-	-	++	-	-	-	125/121
PR 1255	-	++	++	++	-	-	++	-	-	-	125/121
C 3261	++	++	-	-	-	-	++	-	-	-	126/121
C 3281	-	++	-	++	-	-	++	-	++	-	121/130
C 3380	-	-	-	++	+	+	-	-	++	-	130/ 13
C 3612	-	++	-	++	-	+	-	-	++	-	130/124
C 3617	++	-	-	++	-	+	-	-	++	-	126/130

criteria (Simonsen et al. 1982, Crone and Simonsen, 1987). Both of these test systems were applied to the study of putative repeats.

Table 3 illustrates the mutual non-reactivity of the "subfamilies" 818b and 850b, and likewise of 850a and 1255a, where a and b refer to the 2 segregating haplotypes per family. The non-reactive mixes with stimulation indices around unity are shown in the upper left and lower right boxes, autologous mixes being 1.00 by definition. The highly in-creased stimulation indices in the lower left and upper right quarters prove that the 2 sets of non-stimulatory haplotypes are mutually stimulatory, as was indeed expected, though not taken for granted.

The next step was production of haplotype-specific alloantisera. The general design here was cross-immunizing between the a) and b) subfamilies within each family. In each combination, some animals received purified white blood cells (WBC),while others received purified red blood cells (RBC) from their sibling donors. The antisera resulting from in-jection of WBC formed the raw material for preparation of specific anti-F reagents, while antisera to whole blood were the raw material for anti-G reagents. The rationale here was of course that anti-WBC sera should not be contaminated with anti-G antibodies, since G is not supposedly expressed in WBC. In contrast, antisera to RBC are almost certain to contain antibodies to both F and G antigens, since RBC express both F and G.

Either kind of produced "raw material" was now in need of judicious absorptions in order to yield a product acceptable as a specific anti-F or anti-G reagent. Generally, antibodies to non-B blood group antigens were removed by absorption with pooled blood from all the other siblings belonging to the "subfamily" of the antiserum producer. After such absorption the antiserum must of course react with all siblings of the donor "subfamily". This test being passed, the real difficulties begin: how to select the cells for further absorption that will remove as much cross-reactivity as

TABLE 2 Reactivity of new test sera defining new haplotypes in Danish broilers

B-panel	Origin	Specificity of test sera															
		B121		B123		B124		B125		B126		B127		B129		B130	
		F	G	F	G	F	G	F	G	F	C	F	G	F	G	F	G
121	C	++	++	–	–	–	–	–	–	–	–	–	–	–	–	–	–
123	PR	–	–	++	++	–	–	–	–	–	–	–	–	–	–	–	–
124	C	–	–	–	–	++	++	–	–	–	–	–	–	–	–	–	–
125	PR	–	–	–	–	–	–	++	++	–	–	–	–	–	–	–	–
126	C	–	–	–	–	–	–	–	–	++	++	++	–	–	–	–	–
127	PR	–	–	–	–	–	–	–	–	++	–	–	++	–	–	–	–
129	PR	–	–	–	–	–	–	–	–	–	–	–	–	++	++	–	–
130	C	–	–	–	–	–	–	–	–	–	–	–	–	–	–	++	++
1	IS	–	–	–	–	–	–	–	–	–	–	–	–	–	–	–	–
2	"	–	–	–	–	–	–	–	–	–	–	–	–	–	–	–	–
4	"	–	–	–	–	–	–	–	–	–	–	–	–	–	–	–	–
5	"	–	–	–	–	–	–	–	–	–	–	–	–	–	–	–	–
6	"	–	+	–	–	–	–	–	–	++	++	++	–	–	–	–	–
7	"	–	–	–	++	–	–	–	–	–	–	–	–	–	–	–	–
9	"	–	–	–	–	–	–	–	–	–	+	–	–	–	–	–	–
12	"	–	–	–	–	–	–	–	–	–	–	–	–	++	++	–	–
13	"	–	–	–	–	–	–	–	–	–	–	–	–	++	++	–	–
14	"	–	–	–	–	–	–	–	–	–	–	–	–	–	–	–	–
15	"	–	–	–	–	–	–	–	–	–	–	–	–	–	–	–	–
19	"	–	–	–	–	–	–	–	–	–	–	–	–	–	–	–	–
21	"	–	–	–	–	–	–	–	–	–	–	–	–	–	–	–	–

TABLE 3 One way MLR with peripheral lymphocytes of 8 chickens from 4 cock families, 2 individuals from each.

Responder cells			Stimulator cells								
			818b		850b		850a		1255a		
Family, No.		B-type	1	2	1	2	1	2	1	2	
818b,	1	123/15	1.00	0.86	0.48	0.96	8.24	6.57	9.39	0.93	
" ,	2	"	1.61	1.00	0.98	0.58	30.86	48.28	71.19	9.87	
850b,	1	"	0.78	1.11	1.00	0.52	16.18	22.12	31.76	15.40	
" ,	2	"	1.14	0.92	0.78	1.00	3.97	6.54	10.16	22.80	
850a,	1	125/15	9.84	3.93	13.90	2.33	1.00	1.32	1.73	5.33	
" ,	2	"	14.01	4.26	20.45	4.00	0.49	1.00	0.91	0.51	
1255a,	1	"	17.67	11.12	17.32	6.81	0.97	2.36	1.00	1.31	
" ,	2	"	26.38	8.15	29.58	5.28	1.09	1.25	1.23	1.00	

possible with other haplotypes and yet leave an acceptable titer of haemagglutination for the donor cells. In the case of antisera to RBC there is the added difficulty of removing antibodies to F without removing anti-G. Each anti-serum poses its specific problems of this sort, and this is not the place for detailed descriptions of the procedures followed. However two general remarks are called for:

1) Having established first by haemagglutination with a panel of haplotypes, which ones react with a potential new test serum, it may be useful to submit these to the graft-versus-host inhibition release test as earlier described (Simonsen et al., 1987). This test can assist defining the foreign haplotypes that can be used for absorption without removing the main anti-F reactivity wanted.

2) A final quality test of the putatively specific anti-F and anti-G sera is that an anti-F serum must be able to inhibit GVH splenomegaly provoked with donor lymphocytes of the haplotype reacting with the serum, while an anti-G serum must not (F is expressed in T-lymphocytes, G is not).

Table 2 shows the principal outcome of all this labour. Eight new haplotypes have been isolated, each defined by an anti-F and an anti-G of considerable specificity within the P.R. and C. populations. With two exceptions, each set of antisera reacts in the broilers with the specific haplotype only. The exceptions are the test sera for F126 and F127 which both react equally well with both haplotypes, as well as with B6 of the international standard (IS) haplotypes.

Cross-absorptions of the anti-F sera to B6, B126, and B127 is exhaustive of all reactivity, hence there are no serological grounds for sispecting dissimilarity. It is of interest that B126 and B127 lymphocytes both prove to be compatible in B6/6 embryos (no, or very little splenomegaly). The suggestion is therefore that the L alleles of the B6, B126, and B127 are also very similar, though probably not identical. In contrast, the G alleles of these three haplotypes are mutually clearly different: G126 is very similar to G9, G127 is very similar to a G allele isolated from Red Jungle Fowl (not shown), and G6 is clearly distinct from either.

The test serum for F 129 also reacts with B4, B12, and B13, but these are merely cross-reactions, not signs of identity. Otherwise the new antisera have few or no cross-reactions with the standard haplotypes.

Table 4 introduces three more haplotypes isolated from the population of White Cornish, B131, B132, and B133, and gives additional data for B130, which is a frequent haplotype in that population. They are presented together as B21-like, but the similarity to B21 is decreasing from top to bottom as evidenced by the direct compatibility test in B21/21 embryos (with B15/15 embryos as positive controls). In fact, B132 and B133 are clearly incompatible, while B130 is almost wholly, and B131 is partly compatible. It is interesting that the latter two also have F alleles that react with anti F21 while the former two do not. Hence again, similarity for F correlates with the GVH test which is in our belief predominantly a test of L. In our interpretation, an exceedingly strong linkage disequilibrium between F and L makes it a general rule that serological identity or great similarity between F alleles is associated with a comparable similarity in L, reflected in the direct compatibility test in the embryo. The only clear-cut exception is provided by a pair of haplotypes isolated from Red Jungle Fowl, Bw3 and Bw4, which are indistinguishable at F (and G) but clearly different at L (by indirect immunofluorescence), and they do stimulate each otherin both the GVH test and in MLR.

The two bottom lines in Table 4 show the serological reaction pattern with the same panel of test sera applied to the standard B21 and to Bw1, which is a Red Jungle Fowl haplotype which is also compatible in B 21/21 embryos and which also carries a B21-like F allele, but has a different G allele defined by its own test serum. It is interesting that Bw1 also reacts strongly with the B130 reagents for both F and G, which do not react with the standard B21. The evidence as it stands suggests that B130 of Cornish is extremely similar

54

TABLE 4 Serological typing and graft-versus-host tests of
B-21 like haplotypes in White Cornish (B130 - B133)

Haplotype	Specificity of test sera									*Mean log. embryonic spleen weight in	
	F130	G130	F21	G21	GW1	G12	G126	F13	G13	$B^{21}/21$	$B^{15}/15$
B130	++	++	++	-	++	++	-	-	-	1.32 ± 0.13	1.93 ± 0.28
B131	-	-	++	-	-	-	++	-	-	1.65 ± 0.08	2.17 ± 0.11
B132	++	++	-	-	++	++	-	++	-	1.82 ± 0.16	2.26 ± 0.32
B133	++	+	-	-	-	++	-	-	-	2.19 ± 0.03	2.18 ± 0.19
B21	-	-	++	++	-	-	-	-	-	1.25 ± 0.14	2.10 ± 0.22
Bw1	++	++	++	-	++	-	-	-	-	1.32 ± 0.17	1.99 ± 0.19

* Means of 6 B-homozygous embryo spleens per group. Embryos injected i-v with 0.1 ml citrated whole blood from B-homozygous adult donors and killed 5 days later for determination of the spleen weight.

to the wild type version of B21 from the Jungle Fowl. Perhaps it came with the fighting cocks from India.

Table 5 lists the haplotype frequencies in the two investigated broiler lines, calculated by Bernstein's formula on basis of the phenotyping made possible with the new test sera. We have identified a total of 14 new haplotypes, denominated B121-B134 in accordance with the workshop decision in Innsbruck (Briles et al., 1982) that provided for the use of numbers above 100 for provisional designation of new haplotypes. Together with the rediscovered classical B13, and one or more "silent" haplotypes in P.R. that combined to yield 5 birds of "silent" phenotype in a total of 575 birds, we seem to have a pretty complete coverage of the genotypic B-complex repertoire in these 2 lines; perhaps most complete in the C line, where there was no silent phenotype in a total of 149 birds. Not a single bird in either line phenotyped as if it contained more than 2 haplotypes.

B122, B128, and B134 have not been mentioned previously. B122 is a variant of B123, apparently only differing in that its G allele fails to crossreact with anti-G12, whereas the G allele of B123 does. Likewise, B128 is a variant of B125. While G125 crossreacts with anti G12, G128 crossreacts instead with anti G19. B134 is the least well characterized. It crossreacts with anti-G12 but not with any of the anti-F reagents used. Both B134, B133, B132, and B131 have now been obtained as homozygotes from F2 crosses with White Leghorn and specific antisera to their F and G alleles are being produced for further characterization of these haplotypes.

It is not to be expected that the new haplotypes described here will cover nearly the whole of the repertoire present in broiler lines of P.R. and C. used by the chicken industry in general. But it is to be expected that at least some of them will be common. Preliminary tests of other P.R. and C. lines in use in Denmark indicates that this is also the case.

TABLE 5   Haplotype frequencies in the Danish STRYNØ line 74 of
           White Plymouth Rock (PR) and line 85 of White Cornish
           (C).

B-haplotypes	PR-74	C-85
121	0.1608	0.0951
122	0.0400	0
123	0.2333	0
124	0	0.0517
125	0.1104	0
126	0	0.2271
127	0.1213	0
128	0.0390	0
129	0.1433	0
130	0	0.3444
131	0	0.0878
132	0	0.0101
133	0	0.0552
134	0	0.0376
13	0	0.0841
y	0.0933	0
Total	0.9414	0.9931

Haplotype frequencies calculated as $p = 1 - \sqrt{1 - f}$, where f
is the phenotypic frequency. The frequency of the silent
haplotype, By, is calculated as $p = \sqrt{f}$.

DISEASE ASSOCIATIONS WITH B-HAPLOTYPES

In view of the complexities involved in a meaningful serotyping of B-complex haplotypes, as illustrated by the previous section, it is almost impressive that the typing seems to make sense at all in disease associations. The best known example is the correlation between morbidity and mortality from Marek's disease (M.D.) and certain haplotypes of the B-complex, notably B21 going together with resistance.

There are many reports to this effect since the original observation by Hansen et al. (1967). Some show that selection for resistance, but being in the blind with respect to the B-complex, dramatically can increase the frequency of B21 (Briles, 1977). Gavora et al., 1986). Others show B21-linked, relative resistance to standard doses of injected virus in laboratory experiments (e.g. Briles et al., 1982). The latter report also contains the important finding, that the B21 haplotype-conferred resistance in B-heterozygotes greatly depends on the opposite B-haplotype. Thus B2/B21 heterozygotes were with high significance more resistant than four other B21 heterozygote genotypes of the same cross between 2 purebred Leghorn lines. The findings of Gavora et al. (1986) are in full support of that finding.

However, not all investigations have indicated a protective effect of B21 in MDV-infected chickens. The recent work by Hartmann et al. (1986) is an interesting case in point because they obtained the expected protective effect in one but not in the other of 2 crosses involving the same paternal line that segregated for B2, B13, B14, and B21, and 2 different maternal lines, one being homozygous B2/2, the other B15/15. The protective effect of B21 was found in the progeny of homozygous B2-hens, the lack of B21 protection in the progeny of homozygous B15-hens. One interpretation is therefore that the data confirm the findings above of an inter-allelic synergy between B21 and B2. Alternatively, unidentified non-MHC genes may have played a major role that blurred the B21 effect in the other cross. It is of interest that also coccidiosis caused a high mortality in both crosses, and that also the mortality from coccidiosis was significantly reduced in B21 heterozygotes with B2, but not with B15.

The other important group of oncogenic viruses in chickens are the lymphoid leucosis viruses (LLV) and the related Rous sarcoma viruses. There are several adverse effects on production traits associated with LLV infection which have been reviewed by Spencer (1984). He also cites personal communications from Bacon and Crittenden to the effect that the B-complex may influence the development of immunity to LLV, but I am not aware of published details of this work.

Lamont et al. (1987) have published the possibly first example of bacterial infection under MHC control. With injection of relatively low doses of Pasteurella multocida strain X73 causing fowl cholera, there was a highly significant reduction of mortality in a substrain homozygous for B1 as compared to another substrain homozygous for B19. This difference was reproduced with birds from the segre-gating F2 cross, thus strongly reinforcing the conclusion that the difference was truely linked to the B-complex haplotypes. The serotyping was performed with anti-G antibodies, and the authors claim that the linkage is with B-G, but there is no evidence given to the effect that the 2 haplotypes are identical at F and L (which they are probably not).

I want to mention finally some recent studies of our own which were designed primarily as a definitive test of possible associations between the B15, B19, and B21 haplotypes and egg production in White Leghorn. Preliminary studies (Simonsen et al., 1982a) indicated that these 3 haplotypes, which were predominant in the gene pool of "Scandinavian" population synthesized as a mix of commercial hybrids from seven major international firms, responded differently to selection for egg-laying. B15 seemed to be favoured at the expense of B19, in particular in the Swedish and Norwegian parts of the Scandinavian experiment. The new experiment simply monitored the egg production in 3 new lines established from B-heterozygotes of the "Scandinavian" control line. The 3 lines segregated for 2 haplotypes each:

    Line 81: B15 and B19
    Line 82: B19 and B21
    Line 83: B15 and B21

Each generation was the progeny of heterozygous parents of the previous generation and the lines were propagated and monitored for 4 successive generations, F1-F4.

No significant differences were observed in egg-laying parameters correlated to haplotype (Sørensen et al., to be published in detail). However, considering mortality in lines 82 and 83, there was an overall significantly reduced mortality in the homozygous B21 birds in comparison with the remainder (Table 6). All the birds were vaccinated against MD as day-old chickens, and the deaths were due to a variety of causes including coccidiosis, leukosis, and intestinal and oviduct infections.

It is not impossible that the B21 haplotype is good to have around for more reasons than the famous resistance to Marek's disease. The reason why is a different matter, and the answer is unlikely to be forthcoming soon. Not even the mechanism of the relative resistance it confers to MDV infection is at all understood, except that it appears from a study of spontaneous, recombinant haplotypes between B21 and B19 that F and/or L is the decisive part of the B-complex, not G (Briles et al. 1983   ). It may also be pertinent in that connection, that serology and GVH analysis combined has shown a greater variation in G than in F and L in the "family" of B21-like haplotypes which we have now isolated in Copenhagen (B130 and B131 from Cornish, Bw1 from Red Jungle fowl (Table 4), and 2 unpublished haplotypes, one from "Norwegian" Leghorn and one from "Swedish" Rhode Island Red, which show a G allele distinct from the others). We have preliminary findings from RFLP analysis with our avian class II β probe (unpublished) that the B-21-like haplotypes can in fact be distinguished from one another with respect to L, but we still need to perform a comparison with our F and G probes.

TABLE 6 Total mortality in hens from week 18 to week 68 in four generations of two experimental lines, one bred from B15/21 heterozygotes, the other from B19/21 heterozygotes.

Generation	N	Mortality (%) in		Significance
		B21/21	*Remainder	
F1	219	4.4	8.6	n.s.
F2	306	3.1	2.5	n.s.
F3	401	1.0	1.3	n.s.
F4	451	1.8	9.4	$P < 0.01$
Totals	1377	2.3	5.3	$P < 0.025$

* About 75% of the population of genotypes B15/21, B19/21, B15/15, or B19/19

It seems most unlikely that the significance of B-complex typing for genetic disease resistance, or for other components of heredity of practical interest to the chicken industry, will advance beyond the present state of confusion without much more basic research. In particular, it will be necessary to form research groups that are equipped with the gamut of skills needed to correlate all the existing analytical approaches. No cheap and simple typing method of real use will emerge without prior analysis in depth of what it really tells us. Most of what is now going on in this field is in my opinion too haphazard, but the European community clearly has the potential for doing better.

REFERENCES

Briles, W.E., Stone, H.A., and Cole, R.K. 1977. Marek's disease: Effects of B histocompatibility alloalleles in resistant and susceptible chicken lines. Science, 195, 193-195.

Briles, W.E., Bumstead, N., Ewert, D.L., Gilmour, D.G., Gogusev, J., Hala, K., Koch, C., Longenecker, B.M., Nordskog, A.W., Pink, J.R.L., Schierman, L.W., Simonsen, M., Toivanen, A., Toivanen, P., Vainio, O., and Wick, G. 1982. Nomenclature for chicken Major Histocompatibility (B) Complex. Immunogenetics 15: 441-447.

Briles, W.E., Briles, R., Pollock, D.L., and Pattison, M. 1982a. Marek's disease resistance of B (MHC) heterozygotes in a cross of purebred Leghorn lines. Poultry Sc., 61, 205-211.

Briles, W.E., Briles, R., Taffs, R.E., and Stone, H.E. 1983. Resistence to a malignant lymphoma in chickens is mapped to subregion of major histocompatibility (B) complex. Science, 137, 620.

Crone, M., and Simonsen, M. 1987. Avian major histocompatibility complex. In "Avian Immunology: Basis and Practice". (Eds. A. Toivanen and P. Toivanen). (CRC Press, Boca Raton, Florida). Vol. II, pp. 25-41.

Gavora, J.S., Simonsen, M., Spencer, J.L., Fairfull, R.W., and Gowe, R.S. 1986. Changes in the frequency of major histocompatibility haplotypes in chickens under selection for both high egg production and resistance to Marek's disease. Zeitschrift für Tierzüchtung und Züchtungsbiologie, 103, 218-226.

Hala, K., Vilhelmova, M., and Hartmannova, J. 1976. Probable crossing-over in the B blood group system of chickens. Immunogenetics, 3, 97-103.

Hansen, M.P., Van Zandt, J.N., and Law, G.R.J. 1967. Differences in susceptibility to Marek's disease in chickens carrying two different B blood group alleles. Poultry Sci., 46, 1268.

Hartmann, W., Hala, K., Heil, G., and Krieg, R. 1986. Effect of B blood group genotypes on resistance to Marek's disease in Leghorn crosses. In 7th European Poultry Conference, Paris, 24-26 August 1986. World's Poultry Science Association, Branche Francaise. Vol. 1, pp 216-220.

Hutt, F.B. 1949. Genetics of the Fowl. McCraw-Hill Book Co. New York.

Koch, C., Skjødt, K., Toivanen, A., and Toivanen, P. 1983. New recombinants within the MHC (B-complex) of the chicken. Tissue Antigens, 21, 129-137.

Lamont, S.J., Bolin, C., and Cheville, N. 1987. Genetic resistance to fowl cholera is linked to the major histocompatibility complex. Immunogenetics, 25, 284-289.

Pink, J.R.L., Droege, W., Hala, K., Miggiano, V.C., and Ziegler, A. 1977. A tree-locus model for the chicken major histocompatibility complex. Immunogenetics, 5, 203-216.

Simonsen, M., Hala, K., and Nicolaisen, E.M. 1980. Linkage disequilibrium of MHC genes in the chicken. The B-F and B-G loci. Immunogenetics, 10, 103-112.

Simonsen, M., Crone, M., Koch, C., and Hala, K. 1982. The MHC haplotypes of the chicken. Immunogenetics, 16, 513-532.

Simonsen, M., Kolstad, N., Edfors-Lilja, J., Liljedahl, L.-E., and Sørensen, P. 1982a. Major histocompatibility genes in egg-laying hens. Am.J.Reproduct.Immunol. 2, 148-152.

Simonsen, M. 1988. The MHC of the chicken, genomic structure, gene products, and resistance to oncogenic DNA and RNA viruses. In "Veterinary Immunology" (Eds. B.N. Wilkie and P.E. Shewen). (Elsevier, Amsterdam). pp. 243-253.

Spencer, J.L. 1984. Progress towards eradication of lymphoid leukosis viruses - A review. Avian Path., 13, 599-619.

Skjødt, K., Koch, C., Crone, M., and Simonsen, M. 1985. Analysis of chickens for recombination within the MHC (B-complex). Tissue Antigens, 25, 278-282.

Sørensen, P. 1986. Studium af effekten af selektion for vækst hos slagtekyllinger (With English summary and subtitles). (Landhusholdningsselskabets forlag, Rolighedsvej 26, DK-1958 Frederiksberg C., Denmark). pp. 1-314.

Todd, J.A., Bell, J.I., and McDevitt, H.O. 1987. HLA-DQβ gene contributes to susceptibility and resistance to insulin-dependent diabetes mellitus. Nature, 329, 599-604.

SOME EVIDENCE FOR THE PRESENCE OF AN MHC ANALOGUE IN FISH

E. Egberts, R.J.M. Stet, P. Kaastrup[*], C.N. Pourreau & W.B. van Muiswinkel

Dept. of Experimental Animal Morphology & Cell Biology
Agricultural University
P.O. Box 338, 6700 AH Wageningen, The Netherlands
and
[*] Institute for Experimental Immunology
University of Copenhagen
Norre Allè 71, DK-2100 Copenhagen Ø., Denmark

ABSTRACT

Because of disproportional high mortality rates in fish farming, compared to most other animal production systems, there is a growing need for genetic improvement of disease resistance of fish stocks. Present data on defense mechanisms of fish, in particular with respect to the immune system, disclose many similarities, but also differences, with other vertebrate species. In higher vertebrates disease incidence has been linked to the presence of specific products of the major histocompatibility complex (MHC), a polymorphic genetic system which is involved in the regulation of the immune response. Therefore, it is of interest to analyze fish for the presence of such a system, too. The current report describes some experimental data obtained with gynogenetic carp which in this species suggest the existence of cell surface antigens with MHC class I-like properties.

Disease resistance of fish

Improved disease resistance of fish has gained increased attention because of the recent growth of fish farming. Disease outbreaks in fish culture cause substantial economic losses due to high stocking densities, and limited opportunities for individual prophylactic treatment. However, not all fish are equally susceptible to the deleterious effects of a pathogenic agent. For example, we have found clear strain differences in the kinetics of Aeromonas salmonicida pathogenesis in carp (Figure 1), indicative of a genetic component in the disease resistance of this animal species. Therefore, in order to improve the disease resistance of fish, it is of paramount importance to identify this component, and to establish its relationship to the defense mechanism(s) of fish, such as that provided by the immune system.

The immune system of fish

During the last decades it has become increasingly apparent that teleost fish possess a well developed immune system which bears much resemblance to that of other vertebrates. Thus, lymphoid cells as well as mono-

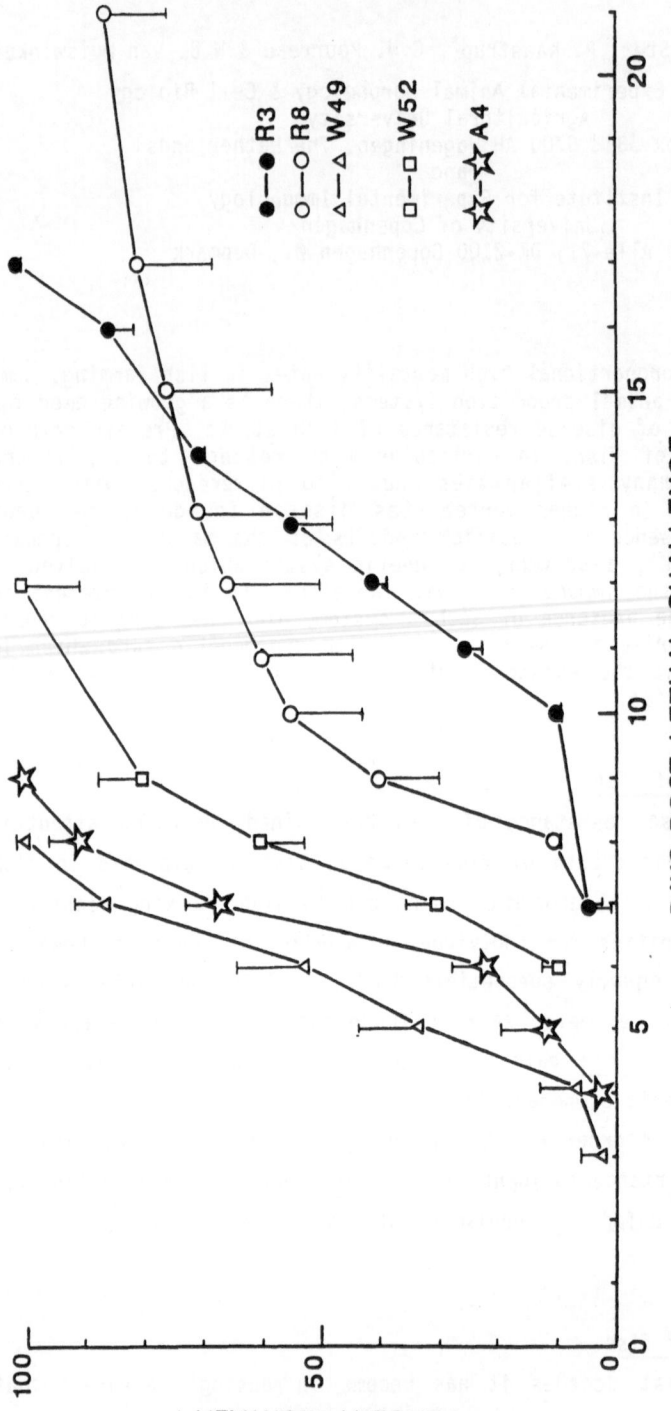

Fig. 1. Five carp strains from different genetic origin were challenged by a subcutaneous injection of a lethal dose of the pathogenic bacterium Aeromonas salmonicida, followed by the registration of the cumulative mortality post infection.
R3 and R8: partially inbred Polish carp strains from the Fish Culture Experimental Station Golysz of the Polish Academy of Sciences (kindly donated by drs J. Szumiec and A. Pilarczyk);
W49, W52 and A4: 1st generation gynogenetic offspring produced in our laboratory from two different females of Dutch (W), and one of Israelian origin (A4), respectively.

nuclear phagocytes and granulocytes of mammals and bony fish share many morphological characteristics, and are in part accumulated in discrete organs like thymus and spleen (Rijkers, 1981). Some of the other lymphoid organs, however, differ: fish lack bone marrow, a bursa and lymph nodes, but instead abundant lymphoid tissue is found in the kidney (pronephros and mesonephros). In addition, no uniformity exists in the histological appearance of the lymphoid organs between different fish species. From a functional point of view, the thymus of bony fish may be considered as a primary lymphoid organ important for the continuous production of lymphoid cells. Pro- and mesonephros may serve the dual purpose of a stem cell compartment, and a peripheral lymphoid organ. The spleen is thought to play a relatively minor role in the immune response of fish (Rijkers et al., 1980).

With respect to non-lymphoid defense mechanisms, there is a large body of evidence that phagocytosis in fish is performed by mononuclear phagocytes (Ellis, 1977). In some species granulocytes take part in this process, too. Several functional analogues of the soluble factors involved in the non-specific mammalian defense mechanisms, like complement, lysozyme, interferon and C-reactive protein, have also been identified in fish (Rijkers, 1982). Furthermore, non-antibody hemagglutinins are present in fish serum which resemble invertebrate agglutinins (Lamers and Van Muiswinkel, 1986).

The capacity in teleost fish to mount a humoral antibody response, and the formation of immunological memory are well developed (Rijkers, 1980). However, here too there are some striking differences in comparison with the mammalian immune system. Although the rate of immunoglobulin synthesis in fish is comparable to that in man, the kinetics of antibody production are very much dependent on ambient temperature. Antibodies in fish can be detected in serum as well as in bile, and mucus of skin and intestine, and are produced by plasma cells which are thought to originate from B-like precursor lymphocytes (Rombout et al., 1986). The latter have only recently been separated from their T-like counterparts by virtue of monoclonal antibodies specific for membrane immunoglobulin (Lobb and Clem, 1982; Secombes et al., 1983; DeLuca et al., 1983). The only free immunoglobulin molecule found sofar is a tetramer whose amino acid composition and physico-chemical properties closely resemble mammalian IgM (Litman, 1977; Marchalonis, 1977). Recent results with the catfish Ictalurus punctatus, however, suggest the presence of different isotypes of the constituent immunoglobulin light and heavy chains within the same individual (Lobb and Olson, 1988). Moreover, the genetic organization of teleost immunoglobulin genes seems to be different from that in mammals (Wilson et al., 1988). The affinity of the fish immunoglobulin molecule for its antigen appears to be relatively low, and only a limited diversity has been reported (Vilain et al., 1984). Nevertheless, from a functional point of view the antibodies are perfectly capable of antigen neutralization.

At the cellular level, bony fish demonstrate many reactivities which are considered to be characteristic of cell mediated immunity, like a mixed leukocyte response (Caspi and Avtalion, 1984; Miller et al., 1986), migration inhibition activity (Jayaraman et al., 1979), delayed type hypersensitivity reactions (Ridgway et al., 1966), and acute allograft rejection (Hildemann, 1970; Rijkers and Van Muiswinkel, 1977). The presence of soluble factors analogous in function with mammalian lymphokines has also been demonstrated (Grondel and Harmsen, 1984; Caspi and Avtalion, 1984).

Considering the central role of the major histocompatibility complex (MHC) in many of the humoral and cellular phenomena of the immune system in other vertebrates, it is reasonable to suppose that fish also dispose of such a system. In eutherian species, an association has been suggested between specific products of the MHC, and disease incidence (Svejgaard et al., 1982), which can be expressed as the relative risk of an individual to contract a particular disease. This finding would also offer a possible explanation for the genetic component of disease resistance in fish which has been observed by us, and others (A. Pilarczyk, pers. communication).

Do fish have an MHC?

Although the theoretical presence of an MHC can be concluded from the above described functional and operational characteristics of the immune system of fish, the formal proof requires a different level of identification, i.e., a serological and biochemical characterization of the gene products of the MHC. An alternative for this approach is the identification of DNA sequences in fish with homology to known heterologous MHC sequences. However, using currently available mammalian MHC DNA probes it has not yet been possible to obtain meaningful hybridization results on teleost genomic DNA (Kaufman, pers. communication; Egberts et al., unpubl. observations). Therefore, serology is needed to initiate the search for the fish MHC analogue. A serological approach has been facilitated using genetically well-defined homozygous fish. Inbred lines are commonly used in mammals, but these are as yet not available for fish, partly due to the long generation time. Instead, gynogenesis in which eggs are fertilized with irradiated sperm, followed by a cold-shock to diploidize the maternal genome, provides a convenient way to obtain families of essentially homozygous carp within a few generations (Nagy et al., 1983).

A first generation gynogenetic carp family was tested for homozygosity using skintransplants between siblings (Van Muiswinkel et al., 1986). These experiments showed that an incomplete homozygosity was achieved for histo-compatibility genes involved in skintransplant rejection. In contrast, two-way mixed leukocyte cultures performed between gynogenetic siblings reveal-ed combinations of siblings identical for MLR genes.

Subsequently, gynogenetic carp siblings were used for the production of alloantisera, with the aim to identify cell surface antigens with MHC-like properties. Hitherto, the siblings were pre-immunized with a skingraft from their prospective peripheral blood leukocyte (PBL) donor, and there-after boosted with PBL. Alloantisera were then tested for their reactivity with donor and third party cells by a direct hemagglutination assay.

Two alloantisera raised between gynogenetic siblings were selected, each defining an allelic specificity designated K1 and K2, respectively (Kaastrup et al., in prep.). Absorption analysis of these alloantisera with erythrocytes expressing the same antigenic specificity showed complete removal of all agglutinating antibodies. This indicates that these anti-K1 and anti-K2 alloantisera were monospecific within the gynogenetic family. Furthermore, immunizations between K-syngeneic gynogenetic siblings did not evoke hemagglutinating antibodies, revealing that the K-antigen is a very strong immunogen present on PBL of carp.

Screening the gynogenetic off-spring (n=23) of a K1/K2 heterozygous female with anti-K1 and anti-K2 alloantisera using a hemagglutination assay resulted in 7 siblings positive for K1 only, 9 positive for both K1 and K2, and 7 positive for K2 only. Similar high frequencies of residual heterozygosity have been observed in histocompatibility studies in goldfish (Nakanishi, 1987). However, a second gynogenetic generation of homozygous K1 as well as K2 females, generated according to methods described by Komen et al. (1988), produced only K1 or K2 homozygous siblings, as demonstrated by hemagglutination assays with anti-K1 and anti-K2 alloanti-sera. These data indicate that the K1 and K2 antigens are products of the same locus, and that the allelic products are expressed co-dominantly.

The cellular distribution of the allelic products K1 and K2 were

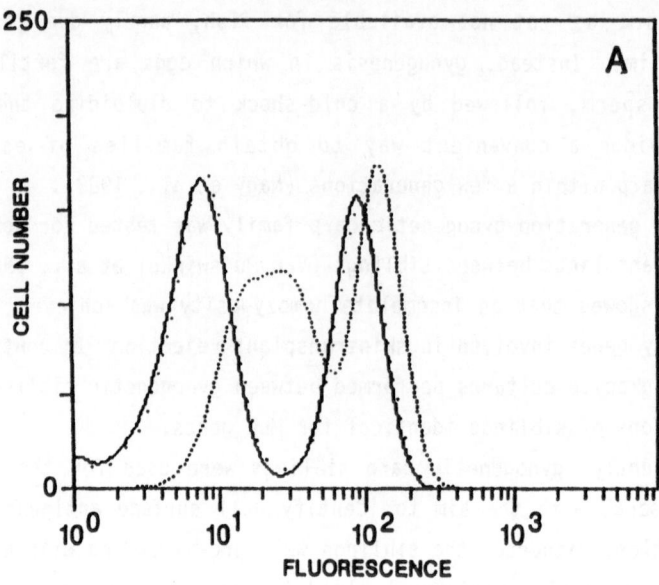

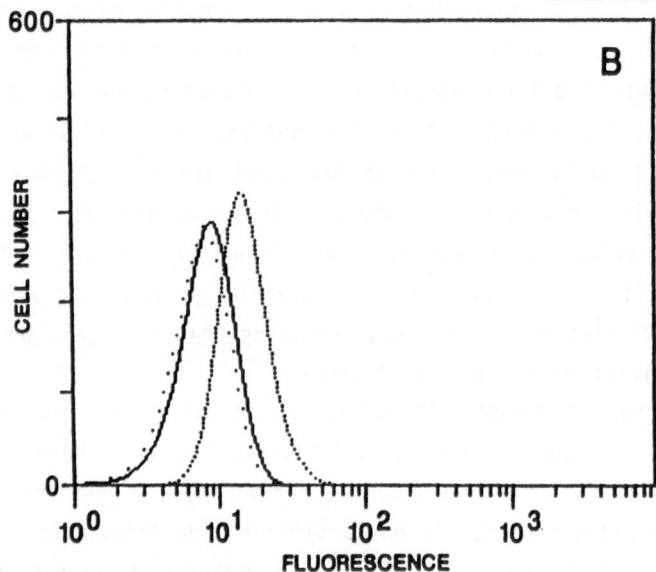

Fig. 2. Flowcytometer (FACS) fluorescence profiles of peripheral blood leukocytes (A) and erythrocytes (B) from a K1-homozygous carp, stained with FITC-conjugated monoclonal antibody WCI12 only (——), or in combination with either anti-K1 alloantiserum (.....) or non-immune carp serum (. . .).

studied using a fluorescence activated cell sorter. Erythrocytes and PBL of both K1 and K2 homozygous siblings were stained with alloantisera. Essentially two populations were revealed in PBL with control non-immune serum and FITC-conjugated monoclonal antibody WCI12 reactive with carp serum IgM (Secombes et al., 1983): an IgM-negative and an IgM-positive population. Staining of PBL with specific alloantiserum showed that the K-antigen is present on both IgM$^-$ and IgM$^+$ leukocytes, as indicated by the increase in fluorescence emmission of both populations (Fig. 2A). Erythrocytes showed a similar increase when stained with alloantiserum (Fig. 2B).

Most antisera to fish erythrocyte antigens that have been examined (cf. Kirpichnikov, 1981) detect blood groups, similar to the ABO blood groups in man. Commonly, these antisera have been produced in mammals and require extensive absorption. However, the blood group antigens detected have never been shown to be present on PBL, which would be a prerequisite for the designation of MHC molecules. The alloantisera described in this study using gynogenetic carp have been shown to detect antigens present on both erythrocytes and PBL. Moreover, K-antigens also seem to be involved in skin graft rejection. Skin grafts exchanged between K-syngeneic carp siblings were rejected significantly slower (MST 15.4 $\pm$ 3.7) compared to transplants between K-(semi-)allogeneic siblings (MST 9.3 $\pm$ 2.5). A similar genetic relationship between erythrocyte antigens and skin graft rejection has been noted in Xenopus laevis (Du Pasquier et al., 1975).

In conclusion, these data indicate that in carp, K-antigens have MHC class I-like properties, because they are present as strong antigens on leukocytes, and behave like histocompatibility antigens in skin graft rejection. The biochemical characteristics of the K-antigens are currently being investigated. It is expected that such an identification and characterization of the fish analogue of an MHC will contribute much to an effective selection of broodstocks for improved disease resistance.

REFERENCES

Caspi, R.R. and Avtalion, R.R. 1984. The mixed leucocyte reaction (MLR) in carp: bidirectional and unidirectional MLR responses. Dev. Comp. Immunol., 8, 631-637.

Caspi, R.R. and Avtalion, R.R. 1984. Evidence for the existence of an IL-2 like lymphocyte growth promoting factor in a bony fish, Cyprinus carpio. Dev. Comp. Immunol., 8, 51-60.

DeLuca, D., Wilson, M. and Warr, G.W. 1983. Lymphocyte heterogeneity in the trout, Salmo gairdneri, defined with monoclonal antibodies to IgM. Eur. J. Immunol., 13, 546-551.

Du Pasquier, L., Chardonnens, X. and Miggiano, V.C. 1975. A major histocompatibility complex in the toad Xenopus laevis (Daudin). Immunogenetics, 1, 482-494.

Ellis, A.E. 1977. The leukocytes of fish; a review. J. Fish Biol., 11, 453-491.

Grondel, J.L. and Harmsen, E.G.M. 1984. Phylogeny of interleukins: growth factors produced by leucocytes of the cyprinid fish, Cyprinus carpio L. Immunology, 52, 477-482.

Hildemann, W.H. 1970. Transplantation immunity in fishes, Agnatha, Chondrichthyes and Osteichthyes. Transpl. Proc., 2, 253-259.

Jayaraman, S., Mohan, M. and Muthukkaruppan, VR. 1979. Relationship between migration inhibition and plaque-forming cell responses to sheep erythrocytes in the teleost, Tilapia mossambica. Dev. Comp. Immunol., 3, 67-76.

Kirpichnikov, V.S. 1981. Genetic Bases of Fish Selection. (Springer Verlag, Berlin).

Komen, J., Duynhouwer, J., Richter, C.J.J. and Huisman, E.A. 1988. Gynogenesis in common carp (Cyprinus carpio L.). I. Effects of genetic manipulation of sexual products and incubation conditions of eggs. Aquaculture, 69, 227-239.

Lamers, C.H.J. and Van Muiswinkel, W.B. 1986. Natural and acquired agglutinins to Aeromonas hydrophila in carp (Cyprinus carpio). Can. J. Fish. Aquat. Sci., 43, 619-624.

Litman, G.W. 1977. Physical properties of immunoglobulins of lower species: a comparison with immunoglobulins of mammals. In "Comparative Immunology" (Ed. J.J. Marchalonis). (Blackwell Sci. Publ., Oxford). pp. 239-275.

Lobb, C.J. and Clem, L.W. 1982. Fish lymphocytes differ in the expression of surface immunoglobulin. Dev Comp. Immunol., 6, 473-479.

Lobb, C.J. and Olson, M.O.J. 1988. Immunoglobulin heavy H chain isotypes in a teleost fish. J. Immunol., 141, 1236-1245.

Marchalonis, J.J. 1977. Immunity in Evolution. (Edward Arnold, London).

Miller, N.W., Deuter, A. and Clem, L.W. 1986. Phylogeny of lymphocyte heterogeneity: the cellular requirements for the mixed leukocyte reaction with channel catfish. Immunology, 59, 123-128.

Nagy, A., Monostory, Z. and Csanyi, V. 1983. Rapid development of the clonal state in successive gynogenetic generations of carp (Cyprinus carpio). Copeia, 1983(3), 745-749.

Nakanishi, T. 1987. Histocompatibility analyses in tetraploids induced from clonal triploid crucian carp and in gynogenetic diploid goldfish. J. Fish Biol., 31, 35-40.

Ridgway, G.J., Hodgins, H.O. and Klontz, G.W. 1966. The immune response in teleosts. In "Phylogeny of Immunity" (Ed. R.T. Smith, P.A. Miescher and R.A. Good). (Univ. Florida Press, Gainesville). pp. 199-207.

Rombout, J.H.W.M., Blok, L.J., Lamers, C.H.J. and Egberts, E. 1986. Immunization of carp (Cyprinus carpio) with a Vibrio anguillarum bacterin: indications for a common mucosal immune system. Dev. Comp. Immunol., 10, 341-351.

Rijkers, G.T. 1981. Introduction to fish immunology. Dev. Comp. Immunol., 5, 527-534.

Rijkers, G.T. 1982. Non-lymphoid defense mechanisms in fish. Dev. Comp. Immunol., 6, 1-13.

Rijkers, G.T., Frederix-Wolters, E.M.H. and Van Muiswinkel, W.B. 1980. The immune system of cyprinid fish. Kinetics and temperature dependence of antibody-producing cells in carp (Cyprinus carpio). Immunology, 41, 91-97.

Rijkers, G.T. and Van Muiswinkel, W.B. 1977. The immune system of cyprinid fish. The development of cellular and humoral responsiveness in the rosy barb (Barbus conchonius). In "Developmental Immunobiology" (Ed. J.B. Solomon and J.D. Horton). (Elsevier North-Holland, Amsterdam). pp. 233-240.

Secombes, C.J., Van Groningen, J.J.M. and Egberts, E. 1983. Separation of lymphocyte subpopulations in carp Cyprinus carpio L. by monoclonal antibodies: immunohistochemical studies. Immunology, 48, 165-175.

Svejgaard, A., Platz, P. and Ryder, L.P. 1982. HLA and disease: a survey. Immunol. Rev., 70, 193-218.

Van Muiswinkel, W.B., Tigchelaar, A.J., Harmsen, E.G.M. and Rijnders, P.M. 1986. The use of artificial gynogenesis in studies of the immune system of carp (Cyprinus carpio L.). Vet. Immunol. Immunopathol., 12, 1-6.

Vilain, C., Wetzel, M.C., Du Pasquier, L. and Charlemagne, J. 1984. Structural and functional analysis of spontaneous anti-nitrophenyl antibodies in three cyprinid fish species; carp (Cyprinus carpio), goldfish (Carassius auratus) and tench (Tinca tinca). Dev. Comp. Immunol., 8, 611-622.

Wilson, M.R., Middleton, D. and Warr, G.W. 1988. Immunoglobulin heavy chain variable region gene evolution: structure and family relationships of two genes and a pseudogene in a teleost fish. Proc. Natl. Acad Sci. USA, 85, 1566-1570.

SESSION 3

MHC polymorphism by protein chemistry and DNA techniques

Chairperson: Dr. R.L. Spooner

SECTION 3

RNA recognition by protein families and DNA techniques

Chairperson: Dr R.E. Spencer

BoLA POLYMORPHISM, BIOCHEMICAL ANALYSIS AT THE PRODUCT LEVEL

E.J. Hensen, I. Joosten, A. van de Poel and M.F. Sanders

Department of Infectious Diseases and Immunology
Faculty of Veterinary Medicine, University of Utrecht
3508 TD Utrecht, the Netherlands

ABSTRACT

The way polymorphism of the MHC is detected, using biochemical analysis at the product level, is fundamentally different from serological, cellular or DNA methods.
Polymorphism of both independent class I and class II loci was investigated separately. A combination of immunoprecipitation with SDS-PAGE gelelectrophoresis and/or isoelectric focusing (IEF) was used. Two major results were achieved.
First the combination of methods made it possible to demonstrate that at least two independent class I loci are expressed for each haplotype. Secondly, the ID-IEF method could be used as a typing procedure. In particular for class II typing, which was not possible until now, this is a new perspective.
Eleven DR-like allelic specificities were found. Therefore, in cattle, studies can now be started to demonstrate the influence of MHC class II in, for example, disease resistance.

INTRODUCTION

Several techniques can be used to define the polymorphism of the BoLA complex. The relevance for a classification of the fine specificities of the products of the BoLA complex depends largely on the influence that the fine specificities will have on the biological function.

The better understanding of the immune interactions in species like men and mice, has illustrated that rather subtle differences in MHC molecules can indeed have tremendous effects on the biological function. Therefore, only a complete and precise classification of variants can answer questions on the effects of variants on the development or the severity of diseases.

The classification of BoLA Class I variants, as defined in the international BoLA workshops, has been based on the results of serology. The results so far are invaluable, but they do not fulfil the requirements to answer the questions on their biological relevance.

Methods have been found which can complement the information obtained by serology and/or can be used to improve the serology.
Cellular techniques, mixed lymphocyte cultures (MLC) and cell mediated lympholysis assays (CML) have been used (Spooner et al., 1987; Teale and

lympholysis assays (CML) have been used (Spooner et al., 1987; Teale and Kemp, 1987; Davies, 1988), but have not resulted in an unambiguous classification. Molecular biology and biochemistry have resulted in new methods of good value. Biochemical methods are used in our studies. These methods are of particular use to define subtle differences between molecules, for which no reagents are available. Such reagents can be prepared once such subtle differences between whole molecules have been detected.

The completeness and discrimination of typing by serology largely depend on the availability of the relevant sera. The typing and analysis by biochemical methods is not based on identification, but on the comparison between molecules based on their overall physiochemical characteristics, such as size and charge. This typing is therefore complete and highly discriminatory.

MATERIAL AND METHODS

Five main steps can be recognised in the procedure, which are listed below, but are described in full detail elsewhere (Neefjes et al., 1986, Joosten et al., 1988a and b).

1). Metabolic labelling of the molecules for identification in the last step. This includes the separation of lymphocytes from a peripheral blood sample and the overnight culture of the lymphocytes with ^{35}S-methionine under standard cell culture conditions. Proteins produced during the culture will incorporate the labelled amino acid at appropriate places.

2). Lysis of the cells containing these molecules. Detergents are used to solubilize the cellmembranes. The membrane proteins, including the MHC molecules are subsequently isolated from the detergent phase.

3). Immunoprecipitation of the molecules of interest with reagents recognising the basic structure of the molecules either class I or class II. Subsequently, the samples are intensively digested by neuraminidase, to remove sialic acid residues. These sialic acid residues are negatively charged and would influence the overall charge of the molecules.

4). Separation of the molecules on the basis of size and/or electric charge in gelelectrophoresis. SDS-PAGE was used for the separation on molecular weight. 1D-IEF (one-dimensional isoelectric focusing) was used for separation on electric charge. Up to 24 samples could be run in parallel on a gel. This enables optimal comparisons.

2D-IEF gels were used if appropriate. First, a separation is done on charge using a rodgelsystem (thin tubes). The content of the tube is layered on top of an SDS-PAGE (along the x-axis) and run in the second dimension (along the y-axis) for a separation on molecular size.

5). Identification of the position of the molecules in the gels by autoradiography and comparison and/or classification of their positions.

RESULTS AND DISCUSSION

Three topics are put forward.

1). The comparison between IEF results and classical serology of class I antigens which has been performed in collaboration with the A.F.R.C. Institute in Edinburgh (Dr. R.L. Spooner c.s.), and was published before (Joosten et al. 1988a).

2). The demonstration of at least two expressed class I locus products for each haplotype in cattle. The results of the serology suggested until now, that only a single class I locus is expressed on the cells for each haplotype in cattle. This was in contrast to all mammalian species investigated. The IEF analysis gave evidence for the expression of at least two class I loci for each haplotype.

3). The description of a DR-like, class II locus. From this locus 11 allelic variants were found, covering all the variants in the population tested (n=98) (Joosten et al. 1988b, in press).

Class I typing

It was shown earlier that the BoLA class I molecules could be precipitated, using reagents that were raised against the human homologous molecules (Hoang Xuan et al., 1982; Bensaid et al., 1988). In the first series of experiments a monoclonal antibody W6/32 (Parham et al., 1979) was used. This antibody was known to be reactive with several mammalian species both from our own experience and others. From the 14 well-defined BoLA serotypes analysed in our study, 12 showed characteristic IEF banding patterns. Using, the W6/32 antibody two BoLA types could not be detected.

But, also 4 new patterns were found in haplotypes which could not be defined by workshop sera. The definition of serological blanks is one of the important advantages of the IEF technique.

Another advantage of the technique will be the subdivision of certain serotypes in (also functional) subtypes or splits. Sofar one good example is under study.

Two or more class I loci

Recently, there have been a few reports that more than one class I locus would be expressed in cattle (Bensaid et al., 1988; Bull et al., 1988). The IEF technique was supposed to give a direct answer. Nevertheless, it took us some time to obtain the evidence. The first complication was a not constant number of bands for each individual serotype. We found that W6/32 did not recognise 2 defined serotypes. This suggested that some allelic products were not recognised. It was hard to envisage in the case of these two serotypes that the products of two loci would both not be recognised, suggesting that indeed only a single class I locus might be expressed. It took some time to realise that under these conditions indeed two independent locus products were both not recognised, and in several serotypes the products of only one locus were detected. When we used more than one reagent in some samples it appeared that these reagents each recognised not necessarily all class I products of a haplotype.

Once these points were clear, one serotype was selected for further study. We knew of this serotype that the use of a second antibody resulted in an extra band after precipitation.

Using the technique of peptide mapping, we could demonstrate that the peptide sequences of the two distinct bands were different. They therefore are the products of independent loci: the results have now been extended to other serotypes (manuscript in preparation). These findings have important implications for the serology of BoLA. A framework for the further development of the serology can now be constructed by the IEF technique.

Class II typing

All methods used thus far had their drawbacks, and no unambiguous classification has been produced so far. Only RFLP data resulted in a classification (Andersson et al., 1986a,b), but until now there was no illustration what proportion of the polymorphism detected would be expressed and therefore functionally relevant.

From the biochemical data it appeared that, with the reagents raised in rabbits against human purified class II products in cattle homologous products were precipitated. In family studies, it became clear, from the IEF patterns that we could detect the allelic products of a single locus. This was selectively the case when the conditions of labelling and the type of detergent were used as described (Joosten et al. 1988b).

The locus products detected are likely DR-like, since there is a striking similarity both with the biochemical characteristics of the products and the reactions with highly specific antihuman DR resp. antisera. From this DR-like locus 11 allelic variants were detected in the 98 animals tested.

TABLE The frequency distribution of the 11 allelic variants which have been found until now. The variants are provisionally named DF (D for D-like and F for focusing)
The data of two halfsib families are included, DF types of the sires are marked *.

BoLA class II DF patterns

Focusing pattern	No. of animals	
	British Friesians	Dutch Friesians
DF1	12	4
DF2	3	11
DF3	16*	4
DF4	20	10
DF5	16	5
DF6	14	11*
DF7	21	22*
DF8	1	--
DF9	4	3
DF10	8*	6
DF11	3	7

(Joosten et al. 1988), 98 animals

This IEF classification can be either used directly in association studies, or as a framework to develop the class II serology. By changing the conditions, the exploration of the DQ-like region, already detected by

DNA studies (Andersson et al., 1986a; Andersson and Rask, 1988) can be started on the product level.

The IEF technique, with the results obtained, has shown to be a potent help to speed up the unravelling of the BoLA polymorphism and, if applied to other species, their MHC polymorphism.

References

Andersson, L., Bohme, J., Rask, L. and Peterson, P.A. (1986a). Genomic hybridization of bovine class II major histocompatibility genes:1. Extensive polymorphism of DQa and DQb genes. Anim. Genet. 17: 95-112.

Andersson, L., Bohme, J., Peterson, P.A. and Rask, L. (1986b). Genomic hybridization of bovine class II major histocompatibility genes: 2. Polymorphism of DR genes and linkage disequilibrium in the DQ-DR region. Anim. Genet. 17: 295-304.

Andersson, L. and Rask, 1. (1988). Characterization of the MHC class II region in cattle. The number of DQ genes varies between haplotypes. Immunogenetics 27: 110-120.

Bensaid, A., Naessens, J., Kemp, S.J., Black, S.J., Shapiro, S.Z. and Teale, A.J. (1988) An immunochemical analysis of class I (BoLA) molecules on the surface of bovine cells. Immunogenetics 27: 139-144

Bull, R.W. et al. (1988). Joint report of the third international Bovine Lymphocyte Antigen (BoLA) Workshop. Anim. Genet., in press.

Davies, C.J. (1988). Immunogenetic characterization of the class II region of the bovine major histocompatibility complex. Ph.D. thesis, Cornell University, Ithaca, New York.

Hoang-Xoan, M., Leveziel, H., Zilber, M.T., Parodi, A.L. and Levi, D. (1982) Immunochemical characterisation of major histocompatibility antigens in cattle. Immunogenetics 15: 207-211.

Joosten, I., Oliver, R.A., Spooner, R., Williams, J.L., Hepkema, B.G., Sanders, M.F. and Hensen, E.J. (1988a). Characterisation of class I bovine lymphocyte antigens (BoLA) by one-dimensional iso-electric focusing. Anim. Genet. 19: 103-113.

Joosten, I., Sanders, M.F., van der Poel, A., Williams, J.L. and Hensen, E.J. (1988b). Biochemically defined polymorphism of bovine MHC class II antigens. Immunogenetics in press.

Neefjes, J.J., Breur-Vriesendorp, B.S., van Seventer, G.A., Ivannyi, P. and Ploegh, H.C. (1986). Definition of new HLA class I subtypes. Human Immunol. 16: 169-181.

Parham, P., Barnstable, C.J., and Bodmer, W. (1979). Use of a monoclonal antibody (W6/32) in structural studies of HLA-A,B,C antigens. Journal of Immunology 123: 342-9.

Spooner, R.L., Innes, E.A., Millar, P., Simpson, P., Webster, J. and Teale, A.J. (1987). Bovine alloreactive cytotoxic cells generated in vitro detect BoLA W6 subgroups. Immunology 61: 85-91.

Teale, A.J. and Kemp, S.J. (1987). A study of BoLA class II antigens with BoT4[+] T-lymphocyte clones. Anim. Genet. 18: 17-28.

COMPARISON of BoLA CLASS I AND CLASS II TYPING METHODS AND THEIR APPLICATION TO MHC FUNCTION STUDIES

J.L. Williams, P. Brown, E.J. Glass, R.A. Oliver and R.L. Spooner

Institute of Animal Physiology and Genetics Research, West Mains Road, Edinburgh, EH9 3JQ.

ABSTRACT

The organisation of the bovine MHC has been investigated by Southern blot analysis using bovine class I and human class II cDNA probes. It is likely that there are multiple class I loci, although only the product of one locus has been identified serologically. The class II region is complex, possibly with similar organisation to the human class II region except that the DP sub-region may be missing. Only DR like products have been identified so far using cross-reactive monoclonal antibodies.

A bovine MHC class I cDNA clone has been isolated and fully sequenced. The amino acid sequence translated from the DNA sequence displays the features expected of classical class I antigens.

Three methods for typing MHC products have been compared, serology, IEF (isoelectric focusing) and RFLP (restriction fragment length polymorphism). Serological and IEF typing correlate for class I alleles, while antigens which are presently not identified by serology can be identified by IEF. Different RFLP patterns can be associated with the same expressed antigen. Therefore for RFLP's to be of value for MHC typing they must first be compared with a method of typing the expressed antigen.

We are using IEF typing for MHC class II function studies, and have shown that the type of response to ovalbumin segregates with class II type in families.

MHC and Disease

Individuals show a remarkable variation in the severity of disease following infection with different pathogenic organisms. There are many factors which influence resistance or susceptibility to disease, however it is the immune system which plays the primary role in controlling an infection once it has been established. The most important event in the induction of an immune response is the recognition of foreign antigen by cells of the immune system. For this recognition to take place the antigen needs to bind to the cell surface products of the major histocompatibility complex (MHC). T cells will only respond to antigen when associated with an individuals own MHC molecules.

There are two classes of MHC molecule, class I molecules are expressed on most cell types and present infection specific antigens, on the surface of infected cells, to cytotoxic T cells. Class II molecules are expressed on antigen presenting cells (e.g. macrophages) which process soluble antigen and present it to T helper cells. MHC molecules are highly polymorphic, with different alleles binding the same antigen with different efficiency and hence affect immune response.

We have therefore concentrated our study of the bovine immune system on the MHC (BoLA).

Genomic Organisation of the MHC

The human MHC (HLA) has been extensively characterised by recombination (Kavanthas et al., 1981, Shaw et al., 1981) deletion mutation analysis (Auffray et al., 1983), cloning (Trowsdale et al., 1984) and pulse field electrophoresis (Hardy et al., 1986). It comprises three major regions, class I, class II and class III and covers about 3000 kbp (see Fig. 1a). The class I region encodes three highly polymorphic classical molecules and in addition there are up to 18 related Qa Tla like genes which display limited polymorphism. The class II region is divided into three sub-regions, DP, DQ and DR, each of which contains at least one pair of expressed α and β genes. The class III region contains genes coding for complement proteins, 21 hydroxylase and tumour necrosis factor and others (for review of the MHC see Klein, 1986).

The organisation of the bovine MHC has been investigated by Southern blot analysis using human and bovine DNA probes (see fig. 1b). Use of a bovine class I cDNA (described below) reveals several polymorphic and non-polymorphic fragments (fig. 3a, Oliver et al., 1989). The polymorphic fragments may arise from classical class I genes while the non-polymorphic fragments arise from conserved genes similar to the Qa Tla genes of the mouse. It is therefore likely that there are multiple classical and non-classical class I genes in the bovine genome.

BoLA class I products were first identified by serology (Spooner et al., 1978) and three international workshops have now identified 34 class I antigens, products of one genetic locus. Expression of other class I loci has been suggested (Stear and Nicholas, 1982, Bensaid et al., 1988a) but this has not been shown conclusively.

The BoLA class II region has been investigated using human locus specific probes, most of these cross-react with bovine genomic DNA

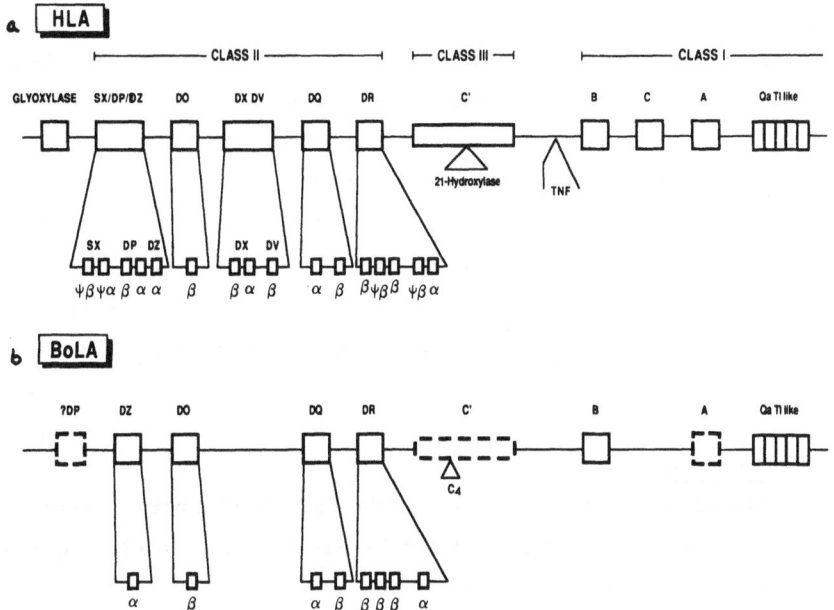

Fig 1. (a) Organisation of the MHC (HLA) region in man (Klein, 1986)
and (b) possible organisation of the MHC (BoLA) region in cattle
(see text).

suggesting the presence of similar genes (see fig. 3b, also Andersson and
Rask, 1988). Thus the bovine genome certainly contains genes similar to
HLA DQ DR DO and DZ. However cross-reaction between probes makes the
presence of a DP like region in cattle uncertain. The only definitive
way to determine the number of genes at each locus is by cloning and
mapping the MHC region. It is however possible to make comparisons of
gene numbers between haplotypes by the analysis of RFLPs. Andersson and
Rask (1988) have shown that the number of DQ like genes is variable
between haplotypes, with single α and β gene in some haplotypes. We
find that there are few fragments revealed using a DRα probe, however a
DRβ probe reveals many fragments (see Fig. 3b). This would be
consistant with a single DRα and multiple DRβ genes, as in man. So
far only DR like molecules have been found at the cell surfaces using
cross-reactive HLA mAbs.

The order of bovine MHC genes on the genome is not known, although
strong linkage between DQ and DR is observed (Andersson et al., 1986;
Sigurdardottir et al., 1988). Field inversion gel electrophoresis (FIGE)

studies have found DQ and DR genes on the same Sfi restriction fragment, demonstrating that they are separated by no more than 1000 kbp (Bensaid et al., 1986b). We find that the linkage between the class I and class II region is less strong. The class I and class II regions have not yet been linked on the genome. From our own cosmid cloning we only find one class I gene per clone. This is similar to the situation in man, whereas in mouse and rat cosmid clones usually contain multiple class I genes (J. Howard, Cambridge personal communication). Therefore it is reasonable to propose a similar organisation for the bovine MHC to that of man (Fig. 1).

BoLA class I genes

We have isolated a bovine class I cDNA (pBoLA-1) from a liver cDNA library (Brown et al., 1988). This cDNA has been completely sequenced, is 1263 bases in length and codes for a protein of 339 amino acid residues. The clone lacks the 5'untranslated sequence, the start codon and some 3'untranslated sequences. Comparison of the sequence of pBoLA-1 with class I sequences from man, mouse, rat etc shows 70-80% similarity in the extra cellular domains, with regions of variation clustered in the protein domains proposed to form the antigen binding site (Fig. 2). The residues present in pBoLA-1 at known variable sites, fit more closely the pattern of substitutions found in classical rather than non-classical (Qa or TLa) class I like genes, suggesting that pBoLA-1 encodes a classical class I molecule.

Identification of MHC alleles

In order to study the role of the MHC in regulating immune response it is first necessary to identify the different MHC alleles. For MHC typing we have used three approaches, serology, restriction fragment length polymorphism and isoelectric focusing of immuno-precipitated MHC products. Serology has been dealt with in detail earlier (Spooner et

Fig. 2. Comparison of pBOLA-1 with human, rabbit and rat MHC class I protein sequences. Maximum alignment has been achieved by introducing gaps (–) where necessary. Clusters of variation are boxed. A mouse (H-2 con) consensus sequence is also shown.

```
First Domain.
                    10          30             50               70              90
pBoLA-1   GSHSMRYFYTAVSRPGLGEPRYLEVGGYVDDTFVRFDSDAPNPRMGPRAPWVEQEGPEYWDQETKAKGTAQTFRANINILGYNQSEA
HLA.CON   g  mr f  svsr  gr  rFIa   df       s aaSp me   ap l qe pe re  qkykaqaqtd vn rtlrg ea
RLA-1            YiS                          AS      NY    EQ MG VE   QI D      V  T  R  A
PD1        P  LS   K   DR DS  FIA            EQ       NY    EQ V IQ Q  RN  E    YGVG  TLR
PD14       P  L   DR      FI   E             EG    rf  YE   EG IQT Q   PM N  IY G RT
H-2.CON    p  l   a   p lge  yis            dte      e  rf  ar m eg    Ere ql gnE .fVd Rtllr   aG
RT1.1      ........L                        Ere  YE        VE       ER     G  DWE I  VD RTL
HM1.6      ...........L  MII W D V  M              LE  VE          M          ER              G

Second Domain.
                    110            130             150             170
pBoLA-1   GSHTFQWMYGCDVGDGRLRGRFMQYYDGRDYIALNEDLRSWTAADTAAQITKRKWEAAGEAERNYLEGCVEWLRRYLETGKDTLLRA
HLA.CON   thqr y   dv p grlL yd yA      K  ian  r   adt a itqr   wea rv eql a legc ewlrrylen etlqra
RLA-1     T  F  E WA  FFH YR A      A          N Q              HA  RE          RE           E  Q
PD1       LS   YL   L LH YR DA      A        M                  D   RS   QL  S              QK  Q
PD14      I I     L L YS ET A       A                           D   RS   QL               QM  Q
H-2.CON   s     rmy  dvg L Ye fA  d c         KT a  maal    rr  Qa a  ryrA          a  r y kn Na  l  T
RT1.1     .....I  IIL S V  H   S  L  YS DA    D        T     T     DL V  HE A      DE  Q              T
HM1.6     .....I II S V  H   S  L YS DA      D        T     T  R                    NR EQ         T

Third Domain.190
                    210            230             250               270
pBoLA-1   DPPKAHVTHHSISGHEVTLRCWALGFYPEDISLTWQRNGEDQTQDMELVETRPSGDGNFQKWAALVVPSGEEQLYTCRVQHEGLQEPLTLKW
HLA.CON   dp  thvT  p1 D  At       a g y aE T  w rD dq  T   A  gtfqk a V     e  R  cHv  g pk ltlrw
RLA-1      L  T  TR PA DR A        AE    D        T      G    T          P        R    P       T
PD1        E T  TR PS DLG          RE    E Q S             T        T    P S      H       R
PD14       s    TR PS DLG          KE    E Q S             T        t    SV  L K .     hvy .     Pe  R
H-2.CON              .Rpe.k         A T   L  eLt  d        a t       SV  L K N   LE           P  QR
RT1.1      G  T   NPEGKT           A  T   K DE EL     M    A    T        VE I     R    HE  S       R
HM1.6                                    A  T         M A       T                                  R

Tm and Cyt Regions.
                    290            310             330
pBoLA-1   ..EPPQPSFLTMGIIVGLVLLVVTGAVVAGVVICMKKRSGEKRGTYIQASSSDSAQGSDVSLTVPKV......
HLA.CON   epssqptipivgivaglavlavvgavvaavmcirrkssgkggsvsqaassdsaqgsdvslta...
RLA-1     EP A TA IV VA VLGVLLIGA VA RRK HS DG G R TP AGGHRD  D MP....
PD1       D A  PVPIV     VL A  M    WR T   G S T AG  D       KDPRV
PD14      D    PVPIV     VL A      WR       G S T AG          KDPRV
H-2.CON   epPpStvsntviiAVlvVLGaaivtGA mkrirnt GlGgd AL PG qtSdl LpDckvmvhdphsl
RT1.1     EPS STD NMETYV YVILGA AII A II AVVAVVMKRRRNTGGKVGVYAPAPSRDSSESSDVSLSDCKA
HM1.6     K  STVPS AV AV   GAVTIIG V AVVR R NTGGKRGNYAQAPDRDSAQ SD SLET..
```

al., this meeting). While it is the most convenient method for typing large numbers of animals it is limited by the sera available. Only the products of one class I locus can be identified, while the BoLA class II products are serologically unidentified. We have therefore also developed other methods for BoLA typing.

RFLP analysis. BoLA class I sequences have been analysed with a bovine class I cDNA (Brown et al., 1988), revealing between 10-15 invariant and 8-10 polymorphic fragments depending upon the enzyme used. The polymorphic fragments can be assigned to haplotypes on the basis of segregation in families. Particular fragments are found in animals inheriting one haplotype from the bull and different fragments are found in animals inheriting the other haplotype, identified by the segregation of serologically defined class I antigens. Other fragments are clearly inherited from the Dams. In one family we identified a TaqI 7.6 kbp fragment which occurred in offspring inheriting the w10 serological specificity from the bull, however on examining RFLPs from unrelated animals we found that this fragment not only occurred in w10 animals, but also in w11 and w32 animals (see fig. 3a). Of the 16 animals included in this sample two are w11 and both have a 6.5 kbp TaqI fragment, however this is not present in all w11 animals, likewise the 5.9 kbp fragment seen here in animals with a putative 2° locus specificity (w30).

BoLA class II sequences have been examined by using HLA, DQ and DR cDNA probes which hybridise strongly to bovine genomic DNA. When these studies were begun we were unable to identify BoLA class II products, we therefore examined the segregation of class II RFLPs in families, inferring the segregation of class II genes with haplotypes defined by serological typing of class I antigen. We are now able to identify class II products by isoelectric focusing (see below). Thus segregation of BoLA haplotypes in the family group shown in fig. 3b was first followed using the class I sera Ed99 and Ed85. We have now found that the Ed99 haplotype also carries the IEF class II EDF8 type and that

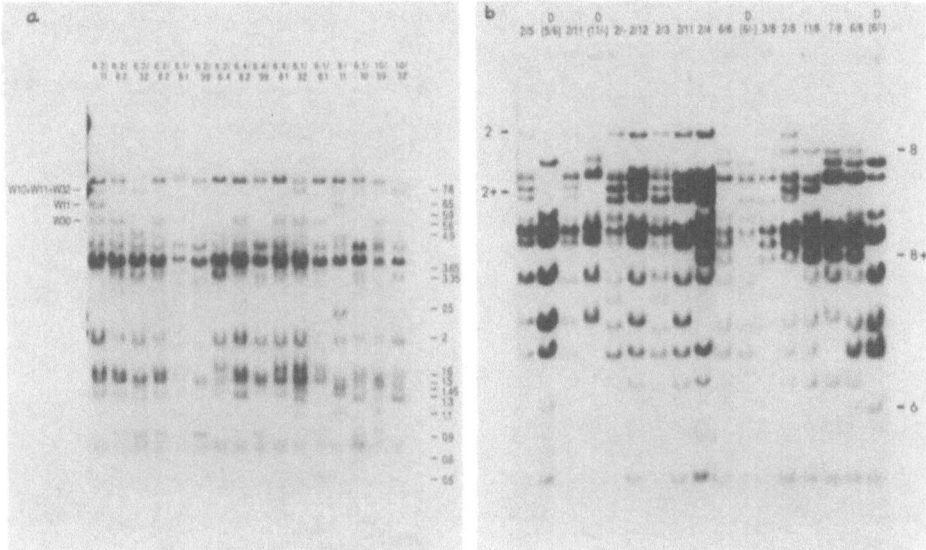

Fig. 3. (a) Genomic DNA from 16 unrelated cattle digested with Taq 1 and
         BoLA class I fragments revealed by hybridisation with pBoLA-1
         (see text).   The serological class I type of the animals is
         shown   at   the   top.   Fragments   occurring   with   particular
         specificities are indicated by arrows. W30 is a putative second
         locus speficity. (b) Genomic DNA from a group of half sibs (the
         bull is class II typed as EDF2/EDF8), digested with Taq 1 and
         hybridised with the β_2 fragment of a DRβ cDNA probe. Four dams
         are included.   Fragments segregating with the bulls MHC class
         II type are indicated by arrows.

of Ed85 the class II EDF2 type.   Different TaqI DRβ restriction
fragments are found segregating with Ed99/EDF8 and Ed85/EDF2 (fig. 3b).
Examination of unrelated EDF8 and EDF2 animals reveals that, as with
class I the same fragments are found in some, but not all EDF8 or EDF2
animals.

The level of polymorphism at the DNA level is greater than that of
the expressed protein hence comparison of MHC class I serological typing
or class II IEF typing with RFLP, shows that the same expressed antigen
can be encoded within sequences which give rise to different RFLPs.
Similar results are found in man (Cohen et al., 1985).

Therefore for RFLP typing to be useful in MHC function or disease
studies it must be correlated with antigen typing. This is best done by
assigning fragments to haplotypes in family studies, then identifying the
different haplotypes expressing specific antigens in population studies.

IEF
We are using isoelectric focusing of immunoprecipitated, ^{35}S labelled
BoLA products as a method of identifying different expressed products.
The patterns seen for class I are complicated. Serologically identified
specificities give rise to between 2 and 6 bands depending on specificity

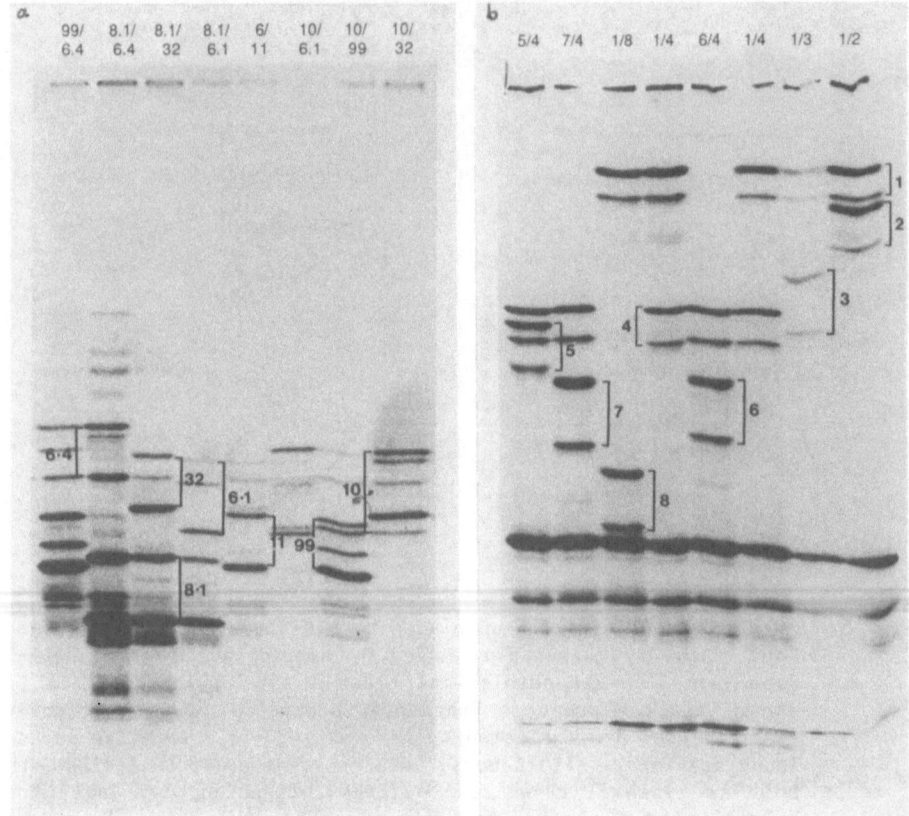

Fig. 4. (a) Autoradiograph of an IEF gel of W6/32 precipitated ^{35}S
labelled BoLA class I molecules from unrelated animals. The
serological class I types are shown at the top and bands
correlating with these types are indicated. (b) BoLA class II
antigens precipitated with polyclonal anti-HLA class II. The
animals were used were the progeny of a bull which was
heterozygous for the bands 1 and 4. Two bands are precipitated
per haplotype. Eight haplotypes are shown and the deduced
typings are at the top.

and monoclonal antibody used for precipitation (see Fig. 4, Joosten et
al., 1988 and Oliver et al., 1989). Nevertheless the bands precipitated
by the monoclonal antibody W6/32 show excellent correlation with
serological specificities, with the advantage that antigens as yet
undefined by serology can be identified.

This is the only method available for identifying BoLA class II
antigens. Comparison of products precipitated by a polyclonal

anti-class II serum with locus specific HLA monoclonal antibodies suggests that the most highly expressed BoLA class II antigen, that we identify by IEF, is analogous to the HLA DR product. The patterns seen on BoLA class II IEF gels are less complex than those for class I, for most animals two major bands are observed (see fig. 4b). It is not known whether the two bands are products of different genes or, as is more likely, the same gene product at different stages of processing or modification. We are currently able to identify BoLA class II haplotypes by IEF and use this method for identifying animals of particular MHC type for functional studies.

MHC and Immune function in cattle

An immune response is triggered by the recognition of antigen in association with MHC class II molecules by T helper cells. This recognition results in the clonal proliferation of T cells and is dependent on the antigen, MHC and T cell receptor molecules involved. We have developed an in vitro T cell proliferation assay to investigate immune response to antigen in cattle (Glass and Spooner, 1988). This assay has demonstrated that T cell proliferation is dependent upon antigen presenting cells and can be inhibited by anti-MHC class II antibodies. Following immunisation with ovalbumin cattle can be divided into two groups, (Glass, Oliver and Spooner, unpublished) responders which show high proliferative response in this assay, and non-responders. Animals with different MHC class II haplotypes show different responses and the type of response segregates with class II type in families.

Conclusions

The MHC of cattle is multigenic with two linked regions encoding the BoLA class I and class II surface glycoproteins. It is likely that there are at least two expressed class I loci, as in man, although only the products of one locus have been positively identified. The class II region contains similar genes to those of the human class II region, although DP like genes may be absent.

BoLA class I alleles have been identified by serology, IEF and RFLP, and class II alleles by IEF and RFLP. For MHC typing serology is the most straightforward method, in the absence of sera IEF typing identifies the expressed products. Polymorphism at DNA level is greater than that of the expressed protein, therefore interpretation of RFLP patterns requires comparison with serology or IEF. The IEF method is the only way of identifying BoLA class II alleles, and we are currently using IEF to identify animals of particular class II types for functional studies.

References

Andersson, L., Bohne, J., Peterson, P.A. and Rask, L. Animal Genetics, 17, 145-204.

Andersson, L. and Rask, L. (1988). Immunogenetics, 27 : 110-120.

Auffray, C., Kuo, J., DeMars, R. and Strominger, J.L. (1983). Nature 304, 174-177.

Bensaid, A., Naessens, J., Kemp, S.J.., Black, S.J., Shapiro, S.Z. and Teale, A.J. (1988a). Immunogenetics, 27, 139-144.

Bensaid, A., Young, J., Kaushal, A. and Teale, A.J. (1988b). Animal Genetics, 19, Sup. 1, 42-43.

Brown, P., Spooner, R.L. and Clark, A.J. (1988). Immunogenetics. (In press).

Cohen, D., Pascale, P., LeGall, I., Marcadet, A., Font, M.P., Cohen-Haguenauer, O., Sayagh, B., Cann, H., Lalouel, J-M and Dawsett, J. (1985). Immunological Review, 85, 87-105.

Glass, E.J. and Spooner, R.L. (1988). Animal Genetics, 19. Suppl. 74-77.

Hardy, D.A., Bell, J.I., Lony, E.O., Lindsten, T. and McDevitt, H.O. (1986). Nature, 323, 453-455.

Joosten, I., Oliver, R.A., Spooner, R.L., Williams, J.L., Hepkema, B., Sanders, M.F. and Hensen, E.J. (1988). Animal Genetics, 19, 103-113.

Klein, J. (1986). Natural History of the MHC. Publ. Wiley and Sons, New York.

Kavanthas, P., DeMars, R., Bach, F.H., Shaw, S. (1981). Nature 294, 747-749.

Oliver, R.A., Brown, P., Spooner, R.L., Joosten, I. and Williams, J.L. (1989). Animal Genetics. (In press).

Shaw, S., Kavanthas, P., Pollack, M.S., Charmont, B. and Mawus, C. (1981). Nature 293, 745-747.

Sigurdardottir, S.. Lundon, A. and Andersson, L. (1988). Animal Genetics, 19, 133-150.

Spooner, R.L., Leveziel, H., Grosclaude, F., Oliver, R.A. and Vaiman, M. (1978). J. Immunogenetics, 5, 335-346.

Stear, M.J. and Nicholas, F.W. (1982). Tissue Antigens, 20, 289-299.

Trowsdale, J., Kelly, A., Lee, J., Carson, S., Austin, P. and Travers, P. (1984). Cell, 38, 241-244.

THE MOLECULAR GENETICS OF THE SLA COMPLEX

Patrick CHARDON
Laboratoire de radiobiologie appliquée -CEA-INRA
78385-Jouy-en-Josas - FRANCE

ABSTRACT.

Knowledge about the overall organization of the SLA complex, the major histocompatibility complex (MHC) in pig, has greatly progressed in recent years thanks to direct gene mapping, to gene cloning and to the study of restriction fragment length polymorphism (RFLP).
The linkage group of the SLA complex, including the Glyoxalase 1 enzyme and the C and J blood group systems is located near the centromere of chromosome 7. As expected, all three classes of MHC gene families were found to make up the SLA region. The class I gene family consists of no more than 7 or 8 genes split into 3 subgroups including 1) the genes coding for the classical transplantation antigens, 2) a divergent member of the class I MHC gene family, whose expression is not ubiquitous, and 3) at least one pseudogene. Positive and negative regulatory elements are found in the 5' flanking sequences of the expressed genes.
The SLA class II genes have been characterized by RFLP and by analysis of a series of clones isolated from a genomic library derived from a miniature swine genome. There might be numerous class II beta genes, while there is probably only one DR alpha and one DQ alpha genes.
The class III region is composed of one set of a C2 gene, a Bf gene and a C4 gene. Evidence also exists that one 21-hydroxylase gene is located in the SLA region.

INTRODUCTION

Before the advent of molecular techniques, our knowledge of the swine major histocompatibility complex (SLA) was obtained by serological, cellular and immunochemical methods. The occurrence of recombinants definitely established that the SLA complex consists of different sets of genes, classified as class I and class II, according to the structural features of their products, their tissue distribution and their functions. Evidence for further complexity came from the demonstration that class I or conventional transplantation antigens were certainly coded for by at least two distinct loci and possibly three (Vaiman et al. 1979). Similarly, it has been shown that the tissue-restricted class II molecules were coded for by more than one locus (Chardon et al. 1981 ; Osborne et al. 1983). Since the SLA complex is known to affect hemolytic complement activity, it was assumed that a set of class III gene(s) were also linked to the SLA complex (Vaiman et al. 1978).

The recent introduction of molecular techniques in the field of immunogenetics, has substantially increased our knowledge of the organization of the SLA region at the molecular level, even though this knowledge still lags far behind what is known about the human HLA or mouse H-2 complex.

The results obtained by these new techniques have often confirmed previous conclusions, particularly the strong homology of the SLA complex to other mammals' MHC. However, they have also revealed some unusual features that appear to be specific to SLA.

THE SLA LINKAGE GROUP : CHROMOSOMAL ASSIGNEMENT

It has been known for some time that the C and J blood group systems are in the same linkage group as the SLA complex (Hruban et al. 1976). Subsequently, the SLA complex was mapped on chromosome 7, and there seems to be agreement on its location near the centromere region, although there is still uncertainty about whether it is located on the short (p) or long (q) arm (Geffrotin et al. 1984 ; Rabin et al. 1985). On the basis of the quantitative differences in glyoxalase 1 activity in different SLA haplotypes, it has been concluded that the glyoxalase 1 gene is closely linked to SLA (Lie et al. 1985). Finally, a syntheny between nucleoside phosphorylase (NP) and mannose phosphate isomerase (MPI) has been described and assigned to chromosome 7 (Dolf and Stranzinger 1986).

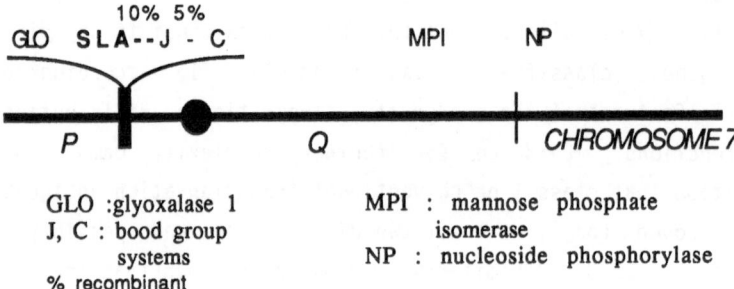

GLO :glyoxalase 1
J, C : bood group systems
% recombinant

MPI : mannose phosphate isomerase
NP : nucleoside phosphorylase

Fig. 1 Organization of the SLA linkage group on chromosome 7.

THE GENETIC ORGANIZATION OF THE SLA CLASS I REGION

The SLA class I gene family, unlike homologous families in other mammals, consists of no more than seven or eight genomic DNA sequences. So far, five of these sequences have been isolated from genome libraries built from the DNA of a homogygous miniature swine with the d haplotype (Satz et al. 1985). Three of these genes, designated PD1, PD14 and PD6 were entirely sequenced.

Structure of SLA class I genes

The PD1, PD14 and PD6 sequences each span about 3.5 kilobases (kb) and appear to correspond to functional but distinct SLA genes. The PD1 and PD14 gene display great homology, although different restriction maps within and around the coding region were obtained. The structure of PD1 and PD14 MHC class I genes, like that of human and mouse class I genes, consists of eight exons separated by introns. The size of the homologous exons of both genes is the same except for the exons of the leader sequence which is 66 bp long in PD1 and 76 pb in PD14. The average percentage of nucleotide homology between the coding sequences of PD1 and PD14 is 88, and for the introns, it is 84. However, for the deduced amino-acid sequences it is about 86. Despite the great resemblance, of the proteins encoded by PD1 and PD14, they are readily distinguisable by serology (Singer et al. 1982).

The 5' upstream region of PD1 contains putative gene regulation sequences. Two positive regulation elements are located within 500 bp of the transcriptional promoter, one of them enhances the expression of the gene in response to interferon stimulation. A negative regulatory element is located 500-1000 bp uptream of the promoter(Singer et al.1987).

The greatest heterogeneity is seen for exon 2, where 28 out of 40 nucleotide differences lead to amino-acid substitution. Most of these differences are clustered in three short stretches of DNA, respectively in nucleotide positions 716-722 (amino-acid residues 40-42), 855-870 (amino-acid residues 66-70) and 883-904 (amino-acid residues 75-82) which all belong to the first extracellular domain of the molecule.

Note that the two last clusters correspond precisely to the hypervariable DNA sequences found within the putative antigen presenting site of an HLA-class I molecule which is clearly endowed with the capacity to accommodate distinct peptides.

The differences observed between exons 3 of PD1 and PD14, corresponding to the second domain, are distributed more regularly along the sequences, although a tendency to clustering is also observed. The homology is greatest for exon 4, which codes for the more proximal third domain and for exons 5 to 8 that encode the transmembrane and intracytoplasmic segments of the molecules, and ranges between 97 and 100 per cent.

Both proteins code for by PD1 and PD14, whose molecular weight is about 45 000 daltons, contain a single potential glycosylation region in the first extracellular domain at amino-acids 86-89, which precisely corresponds to the glyosylation site in HLA-B7. Similarly, both proteins have cystein residues at positions 101-164 and 203-259 that ensure the possibility of constituting stable loops.

The PD6 gene

In contrast to PD1 and PD14, the PD6 sequence only exhibits a loose overall homology with the two other genes, and might infact correspond to a greatly divergent class I gene, while still belonging to the same SLA class I family (Ehrlich et al. 1987).Thus genomic probes specific to PD1 or PD14 genes only hybridize weakly with the PD6 sequence, and the same occurs when PD6-specific probes are hybridized to the other members of the SLA class I family.

The overall homologies of PD6 to PD1 and PD14 are low , around 50% whereas the homology within the exons are markedly greater, reaching 73% The PD6 sequence is common to all pigs and so far displayed no restriction polymorphism.

Judging from the DNA sequence, the PD6 product like other class I antigens, consists of a leader peptide, three extracellular domains, one transmembrane domain and one or more cytoplasmic domains. A potential glycosylation site is found at amino-acid 86-89. One cysteine residue in the second domain is absent, which should preclude disulfide bridge formation, whereas a cysteine residue is found within the transmembrane segment.

It has been shown that four out of five SLA class I sequences cloned, including PD1, PD14 and PD6, can be transcribed and regulated in transfected L cells, but one of them, PD15, appears to be a pseudogene. PD1 and PD14 genes encode class I molecules which are readily identifiable at the surface of mouse transfected cells. So far, similar results have not been obtained for the PD6 product, although a hybrid molecule consisting of exons 1-3 of PD6 and 4 to 8 of H-2 D^d has been found to be expressed in transfected L cells. PD 1 is also expressed in transgenic mice (Frels et al. 1985).

Moderate amounts of PD6-specific RNA were seen in kidney and heart, whereas only a small amount was detectable in thymus and almost none in testis. The liver and peripheral lymphoid organs, i.e. lymph node and spleen express high quantities of PD6-specific RNA, constituting about 10 per cent of the total class I MHC RNA. Among the peripheral lymphoid cells, PD6 is preferentially transcribed in T-cells .

No relationships have yet been established between the SLA class I serologically defined specificities and the SLA sequences already individualized.

Molecular cloning has confirmed the existence of two SLA class I functional genes, but that of the third gene predicted from immunogenetical analyses remains to be shown.

Southern blotting and molecular hybridization have also been carried out in outbred unrelated pigs of the Large White, Landrace and Duroc breeds and in SLA serologically informative families. The number of class I sequences in commercial breeds was the same as in the miniature breed. The results confirmed the molecular SLA class I extensive polymorphism first reported in miniature pig, as few bands were common to any of the distinct haplotypes. As regards serological results, further polymorphism was found at the molecular level (Chardon et al. 1985a, 1985b).

MOLECULAR ANALYSIS OF THE SLA CLASS II REGION

A number of results were obtained from hybridization experiments on enzymatically digested genomic swine DNA using various HLA class II probes specific for HLA-DR and HLA-DQ alpha and beta chains (Chardon et al.1985 a).These results can be summarized as follows : a) after Eco-R1 cleavage of DNA from SLA heterozygotes, the HLA-DR beta probe revealed the existence of about sixteen bands i.e. 6 to 8 genes per haplotype. Many of these bands were polymorphic and when observed in families, always followed the haplotype segregation as revealed by mixed lymphocyte reaction tests (MLR). b) The number of bands obtained by hybridization with a human DQ beta probe never exceeded five or six for a given individual. Usually, one or two intense bands were accompanied by several faint ones, and some of the DQ beta bands appeared to be common to DR beta bands. c) The SLA DQ-like and SLA DR-like heavy chain gene families appeared to be small as no more than 2 bands were observed in any individual. However, with appropriate restriction enzymes, polymorphic fragments were generated. d) Preliminary tests with a specific DP beta probe yielded no clear hybridizing band. The screening of a pig genomic library with a large set of human HLA class II probes has permitted the cloning of a variety of SLA class II sequences. A preliminary analysis of all these clones led to the conclusion that there was one DR alpha non polymorphic gene, one DQ alpha polymorphic gene and numerous class II beta sequences by haploid genome. The class II beta sequences showed extensive cross-hybridization in RFLP tests, therefore precluding the allocation to specific loci (Sachs et al. 1988).

Analysis of several SLA recombinants with all four class II probes confirmed that all the class II genes were clustered inside the SLA region originally defined by the mixed lymphocyte reaction. Furthermore, no alteration in any of the class II bands was observed, which was consistent with the results of the MLR tests (Chardon et al. 1985b).

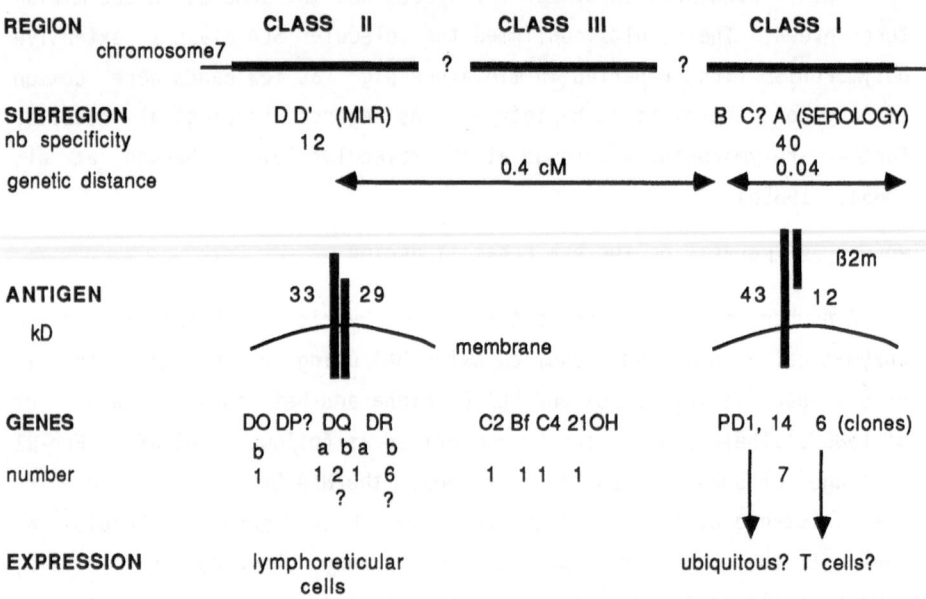

Fig. 2 Genetic organization of the SLA region.

THE SLA CLASS III REGION

Direct evidence for SLA region involvement in complement physiology was obtained when a C4 specific restriction fragment was observed to cosegregate with one SLA haplotype in a family (Kirszenbaum et al. 1985). This was established by Southern blot hybridization using a human cDNA C4 probe of 300 bp, which corresponded to the C4d segment of the protein.

The linkage of C4 to the SLA complex was recently confirmed in miniature swine by american researchers at Ames Iowa, who also demonstrated using C2 and Bf human probes, that the SLA complex contains the C2 and Bf related genes (Lie et al. 1987). Restriction fragment length polymorphism was identified for Bf, which segregated together with specific SLA haplotypes. Although no such polymorphism was found for C2, overlapping restriction fragments were found with the C2 and Bf probes, which strongly suggested that these 2 genes are closely linked.

Similarly, we have recently firmly established that the gene coding for the enzyme 21-hydroxylase or cytochrome P450 also belonged to the SLA complex. Because identical band patterns were obtained by RFLP analysis at the genomic level and by restriction map studies on 21-OH cloned gene there is likely only one of this gene by haploid genome in the pig.

There is as yet no firm indication concerning the relationship firstly, between the different SLA class III gene families, and secondly between the SLA class III region and the class I and class II regions.

FUTURE PROSPECTS

Because the genes coding for tumor necrosis factors (TNF) have been mapped within the human HLA region and mouse H-2 region, we recently started looking for homologous genes in pig. Our interest in 21-OH and possibly in TNF genes was aroused by the role of these genes in important metabolic pathways. Our working hypothesis is that genetically determined variations in both the regulation and structure of these genes might affect performance traits, without entailing harmful, life-threatening consequences.

In conclusion, the SLA region is progressively becoming the best known genomic region in pig, with more than 20 genes already characterized, and many more no doubt remain to be discovered. Accurate knowledge of this important region might radically affect future breeding and selection programmes in pig species.

REFERENCES

Chardon, P., Renard, Ch., and Vaiman, M. 1981. Characterization of class II histocompatibility antigens in pigs. Anim. Blood Grps. and Biochem. Genet. 12, 59-65.

Chardon, P., Vaiman, M., Kirszenbaum, M., Geffrotin, Cl., Renard, Ch. and Cohen, D. 1985. Restriction fragment length polymorphism of the major histocompatibility complex of the pig. Immunogenetics 21, 161-171.

Chardon, P., Renard, Ch., Kirszenbaum, M., Geffrotin, Cl., Cohen, D. and Vaiman, M. 1985. Molecular genetic analyses of the major histocompatibility complex in pig families and recombinants. J. of Immunol. 12, 139-149.

Dolf, G.J., Stranzinger, G. 1986. Pig gene mapping :assignment of the genes for mannosephosphate isomerase (MPI) and nucleoside phosphorylase (NP) to chromosome no.7. Genet. Sel. Evol. 18 , 375-384

Ehrlich, R., Lifshitz, R., Pescovitz, M.D., Rudikoff, S. and Singer, D.S. 1987. Tissue-specific expression and structure of a divergent member of a class I MHC gene family. J. of Immunol. 139,593-602.

Frels, W.I., Bluestone, J.A., Hodes, R.J., Capecchi, M.R. and Singer, D.S. 1985. Expression of a microinjected porcine class I major histocompatibility complex gene in transgenic mice. Sciences. 228,577-580.

Geffrotin, Cl Popescu, C.P., Cribiu, E.P., Boscher, J.,Renard, CH., Chardon, P. and Vaiman, M. 1984. Assignment of MHC in swine to chromosome 7 by in situ hybridizaion and serological typing. Annales de génétique 27,213-219.

Hruban, V., Simon, M., Hradecky, J. and Jilek, F. 1976. Linkage of the pig main histocompatibility complex and J. blood group system. Tissue Antigens 7,267-271.

Kirszenbaum, M., Renard, Ch., Chardon, P., and Vaiman, M. 1985. Evidence for mapping pig C4 gene(s) within the pig major histocompatibility complex (SLA). Anim. Blood Grps. and Biochem. Genet. 16,65-68.

Lie, W.R., Rothschild, M., and Warner, C.M. 1985. Quantitative differences in GLO enzyme levels associated with the MHC of miniature swine. Anim. Blood Grps and Biochm. Genet. 16,243-248.

Lie ,W-R., Rothschild ,M.F., Warner, C.M. 1987. Mapping of C2,Bf,and C4 genes to the swine Major Histocompatibility Complex (Swine Leukocyte Antigen) . J. Immunol 139 ,3388-3395.

Osborne, B.A., Lunney, J.K., Pennington, L., Sachs, D.H.,and Rudikoff,S. 1983. Two-Dimensional gel analysis of swine histocompatibility antigens. J. of Immunol. 31,2939-2944.

Rabin, M., Fries, R., Singer, D., and Ruddle F,H. 1985. Assignment of the porcine major histocompatibility complex to chromosome 7 by in situ hybridization. Cytogenet. Cell. Genet. 39,206-209.

Sachs, D.H., Germana, S., El-Gamil, M.,Gustafsson, K., Hirsch, F.,and Pratt, K.1988. Class II genes of miniature swine.I. Class II gene characterization by RFLP and by isolation from a genomic library. Immunogenetics28,22-29.

Satz, L.M., Wang,L-C., Singer ,D.S,,Rudikoff,S. 1985. Stucture and expression of two porcine genomic clones encoding class I MHC antigens .J. Immunol.135 ,2167-2175.

Singer, D.S., Camerini-otero, R.S., Satz, M.L., Osborne, B., Sachs, D., and Rudikoff, S. 1982. Characterization of a porcine genomic clone encoding a major histocompatibility antigen : expression in mouse L cells. Proc. Natl. Acad. Sci. USA .79,1403-1407.

Singer, D.S., Ehrlich, R., Maguire, J., Golding, H., Satz, L.,Parent,L. and Rudikoff, S:1987. Structure and organization of class I MHC genes in miniature swine. Proc. Intern. Conf. on the Major Histocompatibility Complex of domestic animal Species. Iowa State Press ,in press.

Vaiman, M., Chardon, P., and Renard, Ch. 1979. Genetic organization of the pig SLA complex. Studies on Nine recombinants and biochemical and lysostrip analysis. Immunogenetics 9,353-361.

Vaiman, M., Hauptmann, G.and Mayer, S 1978. Influence of the major histocompatibility complex in the pig (SLA) on serum haemolytic complement levels . J.Immunogenet. 5,59-65.

MOLECULAR ANATOMY OF THE CHICKEN MAJOR HISTOCOMPATIBILTY B COMPLEX

Charles AUFFRAY

Institut d'Embryologie du CNRS et du Collège de France
49bis, Avenue de la Belle Gabrielle - 94130 Nogent s/Marne - France

ABSTRACT

Resistance or susceptibility to the development of tumors induced by Marek's disease virus or Rous sarcoma virus in chickens is controlled by the major histocompatibility B complex. The recent cloning of chicken B complex B-L (class II), B-F (class I) and B-G (class IV) genes has provided new tools for molecular genotyping which should complement available serological and cellular assays and help identifying the B complex genes involved in disease control. The identification of 22 genes in the 320 kb of the B complex which have been cloned make it a prime target for a large scale collaborative sequencing project.

INTRODUCTION

The chicken major histocompatibility (MHC) B complex controls many important immunological functions, as its mammalian homologues. Among them is the control of developping tumors induced by DNA and RNA viruses which are of economical importance (Hala et al., 1981). Thus, resistance or susceptibility to Marek's disease depends on the B complex haplotype. Chickens of the B^{21} haplotype and to a lesser extent those of the B^2, B^6 and B^{14} haplotypes are resistant to infection by the herpes virus causing the disease and the appearance of lymphomas, whereas chickens of other B complex haplotypes are generally more susceptible. Similarly, the ability of tumors induced by Rous sarcoma virus to progress or regress in individual chickens is controlled by several genes, including one linked to the B complex.

Attempts to increase genetic resistance to such diseases are usually based on typing methods for the polymorphic cell surface antigens encoded in the B complex. The most commonly used are haemagglutination tests with polyspecific alloantisera which mostly detect polymorphism of the B-G erythrocyte specific antigens, also called class IV antigens, which have no known equivalents in mammals. Typing for the classical MHC class I (B-F) and class II (B-L) antigens can be achieved by immunofluorescence tests with specific polyclonal or monoclonal antibodies, and by mixed lymphocyte reaction tests for B-L antigens. One primary goal that has been intensively pursued using these methods is to identify which of the B-F, B-L or B-G antigens and genes are responsible for the control of diseases. Many lines of chicken which are highly inbred, homozygous and congenic for the B complex have been established, and used in genetic crosses to search for recombinants within the B complex. In

100

contrast with the situation observed in mammals, the recombination frequency was found to be very low (0.04%). The available recombinants allowed the exclusion of the B-G region, but never separated the B-F and B-L subregions. Only rare recombination events which occurred in the past have been traced by typing outbred populations.

In this paper, I report on the recent progress made towards the identification of B complex genes associated with resistance to viral induced tumors by using recombinant DNA techniques.

CLONING OF B COMPLEX GENES

Our first attempts were based on the use of a human class II β chain DNA probe to detect homologous sequences in chicken by cross-hybridization in low stringency conditions. The HLA-DQβ probe detected mRNA of the appropriate size in chicken tissues where the B-L antigens are expressed, and was used to isolate chicken genomic clones containing B-Lβ genes (Bourlet et al., 1988). Identification of the B-Lβ genes was based on the pattern of transcription of the genes detected by Northern blot analysis, which correlated with the expression of B-L antigens detected with an anti-B-L monoclonal antibody. Probes derived from the isolated genes detected restriction enzyme length polymorphisms (RFLP) between congenic lines of chickens, mapping them to the B complex. Finally DNA sequence analysis of one B-Lβ gene established a high degree of similarity with the mammalian MHC class II β chain and gene (45% at the aminoacid level, and 65% at the nucleotide level).

Subsequently, we have constructed a cosmid library from DNA of the B^{12} haplotype and isolated five genes termed B-LβI to B-LβV. In search of the B-Lα gene(s), we used radio labelled cDNA transcribed from mRNA of various tissues and cell lines as probes to hybridize Southern blots of cosmid clones containing the B-Lβ genes and new overlapping cosmid clones obtained by chromosome walking. The presence of fifteen additional genes was revealed by this method in the 320 kb of DNA that we have isolated and mapped in detail (Guillemot et al., 1988). Two of these genes represent the centromeric rDNA units of the nucleolar organizer located on the same microchromosome as the B complex. By cDNA cloning and DNA sequencing, we identified a family of six class I B-F genes representing the full complement of class I highy related genes in the B^{12} haplotype. The remaining seven genes do not have characteristics of B-Lα or class III complement genes and remain to be identified.

Although much remain to be done to establish a complete molecular map of the B complex, it is already clear that the B-Lβ and B-F genes are very closely linked at distances of 10-30 kb, accounting for the absence of detectable recombination separating them. In addition, the classical B complex genes are found intermingled with unknown genes, a peculiar feature of the B complex not observed in mammals. By immunoscreening of cDNA expression

libraries, cDNA clones for B-G and B-F antigens have been isolated by other laboratories
(Goto et al., 1988 ; J. Kaufman and K. Skjoedt, personal communication). By using these
B-G probes, we have mapped the first B-G gene 15 kb upstream of the B-LβI gene, in a
region that probably represents the centromeric part of the B complex (Fig.1).

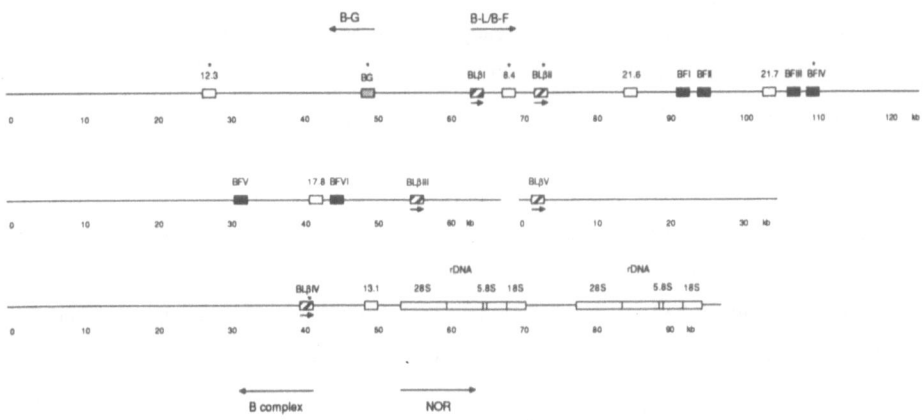

Figure 1 - A molecular map of the chicken B complex
Class I (B-F) genes are indicated by dark boxes, class II (B-Lβ) genes by striped boxes,
B-G by a grey box and other genes by open boxes. Adapted from Guillemot et al. (1988).

MOLECULAR GENOTYPING USING B COMPLEX DNA PROBES

As pointed out above, one of the first evidence that we had cloned B-Lβ sequences came
from the detection of RFLP markers between congenic lines of chickens. Similarly, B-G
probes have revealed extensive polymorphism at the DNA level, and together with the study
of recombinant haplotypes, has allowed to map the hybridizing sequences to the B-G
subregion (Goto et al., 1988).

We have used B-Lβ and B-F probes to search for recombinants in the progeny of crosses
between the CB (B^{12}) and CC (B^4) congenic lines. The results obtained show that there is
complete agreement between serological, MLR and RFLP typing, and no recombinants were

observed. Together with previous studies, this indicates that the distance between the B-F and B-L subregions is below 0.01 cM, in agreement with the physical distance obtained by DNA cloning (Hala et al., 1988). Moreover, this study demonstrate that RFLP markers distinguishing the B[4] and B[12] haplotypes segregate in the F1 and F2 generations in a mendelian fashion together with the B complex.

RFLP analysis has now been extended to the use of B-F, B-L and B-G probes in several animals in a variety of B complex haplotypes in a collaborative effort with laboratories in France (F. Coudert and G. Dambrine), Austria (K. Hala), Switzerland (O. Vassila), Norway (I. Olsaker) and the United States (S. Lamont, C. Warner, S. Bloom and M. Miller). Whereas these studies are for the most part at an early stage, there are already good indications that molecular genotyping of the B complex will develop radily as a powerful tool to complement serological and functional studies in a similar fashion as what was achieved recently in the HLA system. There are already close to one hundred RFLP markers which have been defined and this number is likely to increase rapidely with the use of more restriction enzymes. Patterns of RFLP markers are emerging which are characteristic of each standard B complex haplotype and we have started to describe numerous splits of the B[21] and B[19] haplotypes using B-L and B-F probes (Chaussé et al., 1988).

Molecular genotyping is likely to contribute rapidly to the definition of new markers closely linked to resistance or susceptibility to the development of viral induced tumors, when used in pathological studies. This will possibly help identifying the genes responsible and the mechanism by which they contribute to resistance or susceptibility.

TOWARDS A COMPLETE MAP AND SEQUENCE OF THE B COMPLEX

At present, we have completed a detailed study of 320 kb of DNA of the B complex. When compared with the structure of the HLA and H-2 complexes, the B complex has several original features. The first one is the very small size of the genes determined by the fact that introns are in the range of 100 bp. Thus a B-Lβ or a B-F gene span in the range of 1.5 to 2 kb instead of 5 to 20 kb in mammals. The second is the compaction of the complex of genes itself, with little space between the genes. On the average, there is one gene every 10 kb and in some regions they are even more tightly linked. The third is the absence of clear B-F and B-L regions, the probable absence of class III genes and the presence of numerous unrelated genes in close proximity with B-L or B-F genes. Whereas these features will stimulate much speculation on the evolutionary history of the MHC and the presence of the B complex on an microchromosome, they have obvious practical applications.

One is that a complete molecular map should be obtainable with limited efforts if one assumes a total size of 1000 kb including the B-G region. The current development of new instruments and methods for rapid automated DNA sequencing have been stimulated by the prospect of sequencing the entire human genome. I would like to propose that a one megabase sequencing project aimed at establishing the entire structure of the B complex should be considered as a prime target for evaluating the feasibility of much larger projects and improving instruments and methods for sequence handling and analysis. Moreover, we can already anticipate that a large part of the sequence will be useful to understand the evolution of the MHC class I and class II genes, and will help identifying the structure of the non classical genes and possibly their functions, as well as defining the DNA regions controlling these multiple genes. The establishment of such a large structure should also be regarded as one approach towards the identification of the genes responsible for genetic control of diseases which could then be transferred into the germline of economically important lines of chicken.

Acknowledgments

Work at the Institut d'Embryologie is supported by CNRS, INSERM, INRA and ARC.

REFERENCES

Bourlet, Y., Béhar, G., Guillemot, F., Fréchin, N., Billault, A., Chaussé, A.M., Zoorob, R. and Auffray, C. 1988. Isolation of the chicken major histocompatibility complex class II (B-L) β chain sequences: comparison with mammalian β chains and expression in lymphoid organs. EMBO J., 7, 1031-1039.
Chaussé, A.M., Coudert, F., Dambrine, G., Guillemot, F., Miller, M. and Auffray, C. 1988. Molecular genotyping of four chicken B complex haplotypes with B-Lβ, B-F and B-G probes. Immunogenetics, in press.
Goto, R., Miyada, C.G., Young, S., Wallace, R.B , Abplanalp, H., Bloom, S.E., Briles, W.E. and Miller, M.M. 1988. Isolation of a cDNA clone from the B-G subregion of the chicken histocompatibility (B) complex. Immunogenetics, 27, 102-109.
Guillemot, F., Billault, A., Pourquié, O., Béhar, G., Chaussé, A.M., Zoorob, R., Kreibich, G. and Auffray, C. 1988. A molecular map of the chicken major histocompatibility complex: the class II β genes are closely linked to the class I β genes and the nucleolar organizer. EMBO J., 7, 2775-2785.
Hàla, K., Boyd, R. and Wick, G. 1981. Chicken major histocompatibility complex and disease. Scand. J. Immunol., 14, 607-616.
Hàla, K., Chaussé, A.M., Bourlet, Y., Vassila, O., Hasler, V. and Auffray, C. 1988. Attempts to detect recombination between B-F and B-L genes within the chicken B complex by serological typing, in vitro MLR and RFLP analyses. Immunogenetics, in press.

SESSION 4

MHC and disease associations

Chairpersons: Dr. M. Vaiman and Dr. S. Lazary

POSSIBLE INFLUENCE OF THE CAPRINE LEUCOCYTE ANTIGEN (CLA) SYSTEM ON DEVELOPMENT OF CAPRINE ARTHRITIS ENCEPHALITIS (CAE) IN FAMILY AND POPULATION STUDIES

G. Ruff and S. Lazary
Institute for Animal Breeding, Division of Immunogenetics,
University of Berne, Switzerland

ABSTRACT

The distribution of caprine leucocyte antigens (CLA) in goats from 4 different breeds (N=546) affected by caprine arthritis-encephalitis virus (CAEV) induced arthritis were determined and compared breed for breed with those of infected but clinically healthy controls (N=402)).
Differences in frequencies of some of the CLA specificities between the affected and control groups were found, but after correction of the ordinary P values for number of observed alleles, only the CLA Be7 specificity in the Saanen breed showed a significant deviation at the 0.05 probability level. 11 groups (multiple-case families or halfsibling groups with at least two informative diseased offspring/group), were analyzed for manifestation of the disease and segregation of the parental haplotypes. The results of the maximum likelihood test of association (P<0.005) and the calculated high lod score value of 5.70 gave evidence for linkage between the locus encoding the determined class I CLA alleles and a hypothetic locus(i) coding for genes responsible for arthritis resistance/ susceptibility. The particular class I CLA allele associated with the disease susceptibility varied from family to family, however.

INTRODUCTION

In recent years, 24 codominant alleles encoded by one locus of the goat MHC have been serologically characterized. These caprine leucocyte antigens (CLA) represent class I gene products (Ruff, 1987).

The most ubiquitous infectious disease agent in goats is the caprine arthritis-encephalitis (CAE) virus (Cork et al. 1974; Clements et al. 1979). The virus belongs to the family of Retroviridae, subfamily Lentivirinae. It is closely related to the maedi-visna virus in sheep (Clements et al. 1979; Crawford et al. 1980), as well as to other representatives of the lentivirus subgroup, such as HIV in humans, BIV in cattle and equine infectious anaemia virus (Gonda et al., 1987).

The CAE virus is mainly transmitted through colostrum and milk of infected dams (Cork et al. 1974; Adams et al., 1983) and causes life-long persistent infection (Adams et al. 1980). The clinical manifestation and degree of the disease can vary. Kids are affected by encephalitis between 2 and 4 months of age (Cork et al., 1974). A special form of arthritis is developping in infected animals usually older than 1 year of age. This form of arthritis is not observed in non-infected animals. Interstitial mastitis and pneumonitis are also observed (Crawford and Adams, 1981). Histologically, a mononuclear cell infiltration dominates the chronic inflammatory lesions (McGuire et al., 1986).

The virus occurs worldwide (Adams et al., 1984). The percentage of CAE virus positive goats in Switzerland is high, and is similar to that in Canada and the USA, being about 65-70%. Of the seropositive Swiss goats, 33% show clinical symptoms of arthritis (Krieg and Ruff, 1988).

The aim of the presented investigations was to type CLA alleles in seropositive animals and to determine their distributions in arthritis-affected and arthritis-free groups.

MATERIAL AND METHODS

Animals

The age of the investigated goats ranged from 2 to 10 years, most of them being 2-4 year old females in lactation. 546 animals with clinically diagnosed arthritis were typed for CLA. The following breeds were represented: Saanen (SA; N=146), Toggenburg (TO; N=181), Chamois coloured (CH; N=160) and Appenzell (AP; N=59). The 402 control goats were clinically free of arthritis, but housed together with the diseased animals.

Serological test for CAE virus infection

An indirect enzyme-linked immunosorbent assay (ELISA) was applied using maedi visna virus as antigen (Zwahlen et al., 1983).

CLA-alloantiserum reagents

The preparation, characterization and clustering of the alloantisera have been described in detail (Nesse and Larsen 1987; Ruff 1987). The leader sera (2 to 4) per cluster showing correlation values 0.8 were selected for characterization of single CLA specificities.

Evaluation of the results

The typing results were computed for each CLA specificity and breed. The following parameters were calculated: relative risk (RR); Chi-square (X^2); and the etiological fraction (EF). P values were corrected as follows: P x number of specificities occurring in the breed = cP. For the evaluation of associations in the halfsibling groups, we considered only informative offspring. The following criteria must be fulfilled:

-The common ancestor must be CLA heterozygous.

-At least 2 affected and 1 arthritis-free halfsibling of comparable age and enviroment exist.

-The other parent of the individual must be free of arthritis, or in the case of deceased animals, stated by the owner to have been free of symptoms.

For the statistical analysis, the maximum likelihood test of association as described by Morton (1982) was applied. The family data were also analysed by sequential tests based on the "lod score" method according to Morton (1955). For the tests, the determined CLA locus was considered as the first locus and the susceptibility to disease was treated as a single second "locus". The calculations were carried out on the basis of unknown phases of inheritance.

RESULTS

Population studies

Differences in frequencies of some of the CLA specificities between the affected and control groups were found but after correction of the ordinary P values for number of observed alleles, only the CLA Be7 specificity in the Saanen breed showed a significant deviation at the 0.05 probability level (Ruff and Lazary, 1988).

Multiple-case half sibling groups

Eleven groups (multiple-case families or halfsibling groups with at least two informative diseased offspring/group), were analyzed for manifestation of the disease and segregation of the parental haplotypes.

Figure 1 shows the pedigree of halfsiblings sired by a Saanen buck. He was not affected by arthritis during his use and transmitted either Be2 or Be7 CLA specificities to his offspring. All the animals in this group were bred and held at the same farm. Of the 7 descendants investigated, 3 suffered from CAE virus-induced arthritis. All of the cases inherited the antigen Be2 from the sire, whereas non of the arthritis-free animals expressed the paternal Be2 specificity. The ratio of Be2:Be7 was 3:0 in the affected group and 0:4 in the nonaffected group, respectively.

Another Toggenburg halfsibling group is shown in Figure 2. The clinically ill sire transmitted either Be9 or Be20 specificity to his 9 offspring. 3 of the 4 affected animals carried the Be9 antigen, in contrast to only 1 of the non-affected goats. The animals were all bred on the same farm and infected with colostrum of seropositive dams from the first day of life. The ratio of Be9 to Be20 was 3:1 in the affected group and 1:4 in the arthritis-free animals.

In order to show a genetic association between the MHC region and the susceptibility to arthritis, the data has been subjected to the maximum likelihood test of association as described by Morton (1982). The combined relative risk $\aleph$ turned out to be 3.16. The combined test for association resulted in a χ^2_1 value of 41.12 (P$<$0.0001). The test for heterogeneity among m investigated samples (m=11) gave a χ^2_{m-1} of 20.04 (P$<$0.025). Finally Fischer's F test was performed on the above χ^2 values and showed a significant level of P$<$0.005 with F=20.52.

The calculated high lod score value of 5.70 gave evidence for linkage between the locus encoding the determined class I CLA alleles and a hypothetic locus(i) coding for genes responsible for arthritis resistance/ susceptibility. The particular class I CLA allele associated with the disease susceptibility varied from family to family, however.

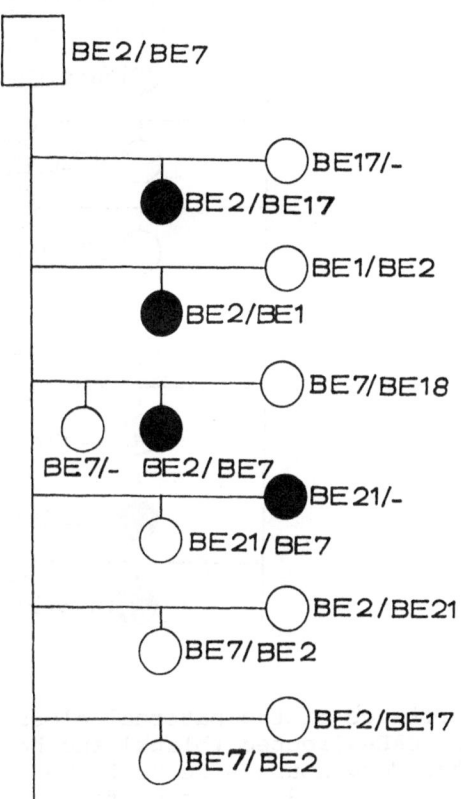

Fig.1   Inheritance of the paternal class I-marked haplotypes in CAE-affected (black) and healthy (white) progeny of the Saanen sire. The ratio of the antigens Be2:Be7 was 3:0 in the informative affected and 0:4 in the healthy group. Squares symbolize male and circles female goats.

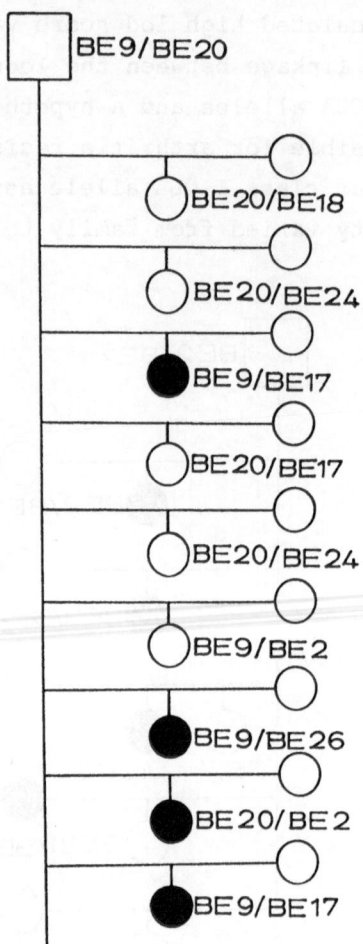

Fig.2   Inheritance of the paternal class I-marked
haplotypes in CAE-affected (black) and healthy (white)
progeny of the Toggenburg sire no.2. The ratio of the
antigens Be9:Be20 was 3:1 in the affected and 1:4 in the
healthy group. Squares symbolize male and circle female
goats.

CONCLUSION

These data provide the first evidence that CAE virus induced arthritis in the goat is genetically influenced by the MHC system, but also suggest that susceptibility/ resistance genes are not directly associated with the determined class I gene products but are rather in close genetic linkage.

ACKNOWLEDGEMENT

This work was supported by the Swiss Natiolfonds for Scientific Research, grant 3.879-0.88

REFERENCES

Adams, D.S., Crawford, T.B. and Klevjer-Anderson, P. 1980. A pathogenetic study of the early connective tissue lesions of viral caprine arthritis encephalitis. Am. J. Path. 99:257-278.

Adams, D.S., Klevjer-Anderson, P., Carlson, J.L., McGuire, T.C., and Gorham, J.R. 1983. Transmission and control of caprine arthritis-encephalitis virus. Am. J. Vet. Res. 44:1670-1675.

Adams, D.S., Oliver, R.E., Ameghino, E., DeMartini, J.C., Verwoerd D.W., Houwers, D.J., Waghela, S., Gorham, J.R., Hyllseth B., Dawson, M., Trigo, F.J., and McGuire, T.C. 1984. Global survey of serological evidence of caprine arthritis-encephalitis virus infection. Vet. Rec. 115:493-495.

Clements, J.E., Narayan, O., and Cork, L.C. 1979. Biochemical characterization of the virus causing leukoencephalitis and arthritis in goats. J. Gen. Virol. 50:423-427.

Cork, L.C., Hadlow, W.J., Crawford, T.B., Gorham, J.R., and Piper, R.C. 1974. Infectious leukoencephalomyelitis of young goats. J. Infect. Dis. 129:134-141.

Crawford, T.B., Adams, D.S., Cheevers, W.P., and Cork, L.C. 1980. Chronic arthritis in goats caused by a retrovirus. Sience 207: 997-999.

Crawford, T.B. and Adams, D.S. 1981. Caprine arthritis-encephalitis: clinical features and presence of antibody in selected goat populations. JAVMA 178:713-719.

Gonda, M.A., Brown, M.J., Carter, S.G., Kost, T.A., Bess, J.W., Arthur L.O., and Van Der Maaten, M.J. 1987. Characterization and molecular cloning of a bovine lentivirus related to human immunodeficiency virus. Nature 330:388-391.

Krieg, T. and Ruff, G. 1988. CAE : Caprine Arthritis-Encephalitis. Der Kleinviehzuchter 3:93-98.

McGuire, T.C., Adams, D.S., Johnson, G.C., Klevejer-Anderson, P., Barbee, D.D., and Gorham, J.R. 1986. Acute arthritis in caprine arthritis-encephalitis virus challenge exposure of vaccinated or persistently infected goats. Am. J. Vet. Res. 47:537-540.

Morton, N.E. 1955. Sequential tests for the detection of linkage. Am. J. Human Genet 7:277-318.

Morton, N.E. 1982. Outline of genetic epidemiology. pp 95. S. Karger, Basel, New York.

Nesse, L.L. and Larsen, H.J. 1987. Lymphocyte antigens in Norwegian goats: serological and genetic studies. Animal Genetics 18:261-268.

Ruff, G. 1987. Investigations on the Caprine Leucocyte Antigen (CLA) System. Thesis no. 8468, ETH Zurich.

Ruff, G. and Lazary, S. 1988. Evidence for linkage between the caprine leucocyte antigen (CLA) system and susceptibility to CAE virus-induced arthritis in goats. Immunogenetics 28:303-309.

Zwahlen, R., Aeschbacher, M., Balcer, T., Stucki, M., Wyder-Walther, M., Weiss, M., and Steck, F. 1983. Lentivirusinfektionen bei Ziegen mit Carpitis und interstitieller Mastitis. Schweiz. Arch. Tierheilk. 125:281-299.

STATISTICAL ASPECTS OF CATTLE MHC (BOLA) AND DISEASE ASSOCIATIONS
EXEMPLIFIED BY AN INVESTIGATION OF SUBCLINICAL MASTITIS

H. Østergård

Department of Animal Genetics
The Royal Veterinary and Agricultural University, Copenhagen
Bülowsvej 13, DK-1870 Frederiksberg C, Denmark

ABSTRACT

A preliminary investigation of associations between cattle MHC (BoLA) and subclinical mastitis measured by somatic cell counts (SCC) of quarter milk samples of 239 animals from 3 different breeds and crossbreeds is presented. The statistical analysis is based on a statistical model which includes effects of external and genetic relationship groups (fixed or random) as well as effects of MHC genotypes or alleles. With this model it is possible to estimate unweighted allelic effects of all alleles simultaneously by multiple regression. An analysis of variation in cow average logarithmic SCC indicated no significant random sire effect, no significant influence of BoLA alleles but a significant effect of herds, lactation numbers and breeds. These results are discussed in relation to other investigations of MHC associations in cattle concerning mastitis, bovine leukosis and tick and worm infestations. A general problem of demonstrating significant and/or consistent disease associations from population data is discussed in relation to loci studied, disease parameters studied, experimental design and selection on MHC.

INTRODUCTION

In the following are summarized the results obtained from a detailed discussion of statistical models for MHC associations in farm animals (Østergård et al., 1988). The choice of statistical model for studying a potential association between an MHC locus and a disease related trait depends on the available observations. In most experiments environmental conditions vary and this may influence the expression of the trait considerably. Therefore, the statistical model should include the possibility of correcting for effects of external factors, e.g., different herds, different years, etc.. Also a correction for genetic relationship, i.e., for halfsib groups correction for sire effects, should be possible within the model. When the disease parameter is quantitative, an analysis of variance with external and genetic relationship groups defining incomplete blocks for comparison of effects of MHC genotypes and alleles has these characteristics. In addition, additive genetic effects of background genes may be evaluated by considering the genetic effects as random. The special requirement for studies of MHC associations is that the statistical model should enable an evaluation of different hypotheses for the influence of the MHC locus on the variation of the trait. The alleles at the considered MHC locus may influence the trait in an additive or a non-additive way. In the former case the alleles are the relevant entities to study whereas in the latter it is the genotypes. In case a particular allele is a susceptibility gene, it may be relevant to compare the influence of this allele to the average of the remaining alleles but otherwise it is most informative to consider the effect of all alleles simultaneously and from this describe the ranking of the alleles with respect to the variation of the trait. These aspects may be included in the model by defining allelic effects as multiple regression coefficients of variables counting the number of specific alleles in the genotypes (0, 1, or 2). This is designated a multiple allele procedure.

In most models used previously a set of analyses, one for each allele, are necessary for describing the influence of different alleles. Each analysis includes the effect of the MHC locus by means of a single regression coefficient (or class factor) corresponding to presence or absence of a particular allele (single allele procedure). In these models unweighted allelic effects cannot be estimated. In stead, the evaluation of the influence of each allele is by means of a comparison of the average allelic effect of that particular allele, averaged over genotypes including the allele, with an average effect of all other alleles, averaged over genotypes not including the allele. The weights included in the averages differ from sample to sample being complex functions of the observed number of individuals with the different genotypes.

When a dichotomous trait as diseased or not diseased is considered the considerations above are still valid. In this case, an analysis of deviances based on a logistic regression model can be applied as exemplified by Østergård et al. (1988).

STATISTICAL ANALYSIS OF ASSOCIATIONS BETWEEN CATTLE MHC (BOLA) AND SOMATIC CELL COUNTS (SCC)

This study was initiated to analyse further the association between BoLA-allele w16 and high incidence of mastitis indirectly observed by demonstrating a similar association for the M' blood group factor (Larsen et al., 1985). A possible biological explanation for this association may be that the M' factor influences the growth of specific bacteria directly (Kaartinen et al., 1988).

MATERIALS AND METHODS

A sample of 308 cows from 8 herds and of three breeds (Danish Red, Danish Black Pied and Danish Jersey) and crossbreeds were studied. The cows were divided into 3 groups according to lactation number. A subsample of 239 cows with 2 well-defined BoLA-alleles and at least 6 animals positive for each allele was considered for the statistical analysis. The paternity was known for 180 of these cows and they were the daughters of 83 bulls.

The BoLA typing was performed in accordance with the 3rd International Comparison Test (joint report in prep.) using 71 antisera detecting 24 BoLA specificities (alleles at one locus). In the statistical analysis only 18 specificities were included: The old notation was used for the specificities w4, w5, w6, w6.2, w6.4, w7, w8, w9, w10, w11, w12.1, w12.2, w16, w20; the provisional notation for 21, 22 and 28; and CO12.3 was the local specificity CODO43 which seems to be a subtype of w12.

Quarter milk samples were collected 6 times with 14 days intervals. Bacteriology and various indirect inflammation markers were investigated but for this analysis only direct somatic cell counts (SCC) were con- sidered, i.e., the average logarithmic SCC-value for each cow.

The statistical analysis was performed by means of the following general linear model (multiple allele procedure):

$$\overline{\log (SCC)}_{ijkl(s,t)n} = m + a_i + b_j + c_k + (h_s + h_t) + e_{ijkln} \qquad (1)$$

with m = general mean

a_i = fixed effect of ith herd

b_j = fixed effect of jth lactation number group

c_k = fixed effect of kth breed

h_x = fixed effect of xth allele (additive effect)

e_{ijkln} = random effect of nth cow with genotype l consisting of the alleles s and t

All fixed effects were relative to the average, i.e.,

$$\Sigma_i \, a_i = \Sigma_j \, b_j = \Sigma_k \, c_k = \Sigma_x \, h_x = 0.$$

RESULTS AND DISCUSSION

The model Eq. (1) was chosen to describe the effect of herds, lactati- on number groups, breeds and BoLA alleles on the variation in average logarithmic somatic cell counts. The effect of sires was not included in the model since the variance component corresponding to a random sire effect was not significantly different from zero. Interactions between alleles could not be included in the model due to the small sample size. Similarly, interactions between alleles and breeds could not be included so it was assumed that alleles had the same effect in different breeds. This may be unrealistic if the effects of the locus studied were due to gametic disequilibrium to a "disease locus".

The BoLA alleles could not be demonstrated to have a significant influence on somatic cell counts (Table 1), but herds and lactation number groups had a very significant effect. The effect of breeds was unexpectedly small due to confounding of breeds with BoLA alleles. If the effect of breeds was not corrected for effect of BoLA alleles, a very significant effect was found (P=0.0001).

TABLE 1   Analysis of variance for $\overline{\log(SCC)}$

Source	DF	F-value	P-value
herd	7	2.75	.0094
lactation no.	2	20.50	.0001
breed	3	2.41	.0681
BoLA	17	0.66	.8396

The reason why it was not possible to demonstrate an influence of the BoLA alleles on SCC could be that 1) there is no effect, 2) the effect is based on interactions between alleles or 3) the number of animals with the interesting alleles were too small in this sample. However, the allelic effects estimated independently of the sample composition gave valuable information about the ranking of the alleles with respect to SCC. The estimates of allelic effects were shown in Fig. 1 together with their standard errors plotted from the base line to facilitate comparisons. The allelic effects were rather similar for the different alleles indicating that no single "susceptibility or resistance gene" was present. One allele, 21, deviated significantly from the average of the 18 alleles when not correcting for the number of comparisons made (e.g. by using the Bonferoni inequality (cf. Østergård et al., 1988)). The alleles 21 and 28 had nearly identical effects in that they increased the somatic cell counts about 2.5 times compared to the average effect of the 18 alleles. A similar conclusion would not be possible to draw from the single allele procedure where unweighted parameter estimates are not available. The allele w11 had the opposite effect in that it approximately halved SCC compared to the average effect. The effect of w16 was to increase the so-

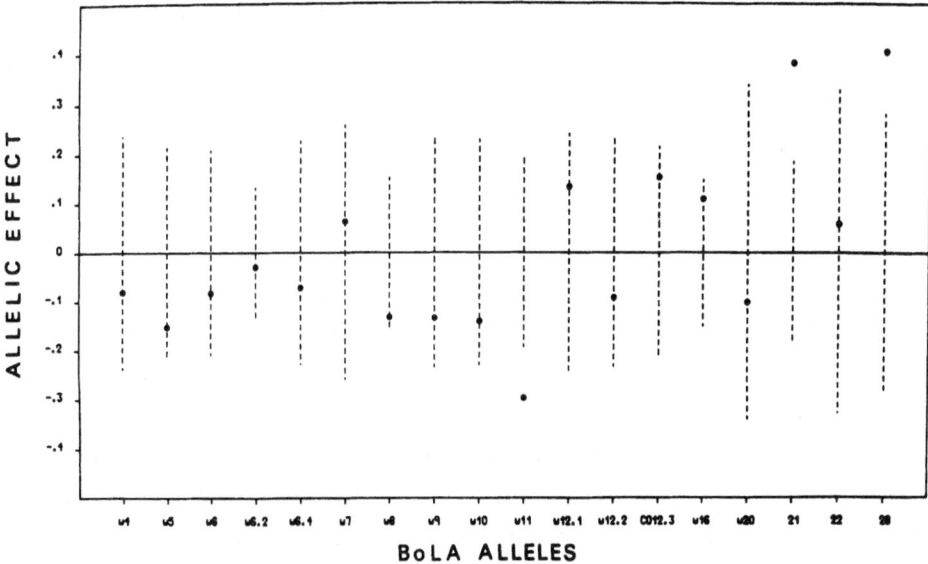

Fig.1   Additive effects of alleles on a logarithmic scale for SCC
(summing to zero). Standard errors indicated by broken lines plotted
from the base line.

matic cell counts about 1.3 times compared to the average effect and so
the effect of w16 was consistent with previous results.

REVIEW OF DISEASE INVESTIGATIONS

Investigations of MHC and disease associations in cattle populations
published over the last few years have considered, beside mastitis, bovine
leukosis and tick and worm infestations, and the effect of one Class I
locus. In the following, mainly statistical aspects of the investigations
are discussed.

The studies of bovine leukosis are the most   detailed. Progression of
bovine leukemia virus infection has been studied in 2 different herds (117
Shorthorn cattle and 240 Holstein-Friesian cattle, respectively) where
also two halfsib groups were considered (Lewin and Bernoco, 1986; Lewin et
al. 1988) and in a group of 88 cattle of Australian Illawarra Shorthorn
sampled according to a case-control scheme (Stear et al., 1988a). Also a
few other animals of various breeds were included in the latter analysis.
In the studies by Lewin and coworkers the parameters considered were

frequency of seropositives, percentage of peripheral blood B-lymphocytes and persistent lymphocytosis whereas in the latter study only persistent lymphocytosis was reported. The number of alleles present in the three populations in reasonable numbers were 10, 9 and 6, respectively. Some alleles were present in all samples but different effects were found using the single allele procedure. One highly significant difference was obtained (P=0.00008) and this corresponds to a very significant test even if the test level was corrected according to the Bonferoni inequality method: w8.1+ animals were more frequently seronegative than w8.1- animals. The possibility of sire effects was not discussed. In the case-control data analysed be Stear and coworkers (1988) the two groups of animals seemed to have rather different allele frequencies (not significant when corrected) and a reason for this may be that it is very difficult to define controls in farm animals. In the analyses of segregation in halfsib families, some significant differences were obtained. In conclusion, aspects of BLV progression seems to be influenced by genes in the MHC region.

Mastitis has been studied in 2 investigations in addition to the present one: 1) The animal material consisted of 64 cows from 3 Icelandic dairy herds (Oddgeirsson et al., 1988) and the parameters based on measurements of quarter milk samples were infected or not infected, direct somatic cell counts and ATP and antitrypsin level. Fifty five sera were used for BoLA typing but the corresponding number of specificities observed in the samples were not shown (w16 was not detected). Only sera with significant associations were mentioned and so it was not possible to calculate the corrected P-values. It was concluded based on the single allele procedure that BoLA had an effect on SCC and antitrypsin level. The effect of sires was not discussed. 2) Breeding values for clinical mastitis for a number of sires and their male offspring (cf. Solbu et al., 1982) were studied (Spooner et al., 1988). Analyses of breeding values have the drawback that they can only demonstrate additive genetic effects. Different statistical models were considered in relation to corrections for sire effects and these models produced somewhat different results. The allele w16 and at least 4 other alleles were included in the analyses. The allele w16 appeared to be associated weakly with susceptibility. In conclusion, in all studies the effect of w16 has been consistent whereas too little information has been gathered with respect to the other alleles.

Investigations of tick and worm infestations have been performed by

Stear and coworkers. One herd (332 Brahman x Shorthorn cattle) was investigated over several years for number of surviving ticks after artificial
infestation (Holroyd and Stear, 1985). In a selection experiment of
Africander-Hereford cross cattle with 6 sire lines selected for high or
low faecal worm egg counts, faecal worm egg counts after natural infection
were measured in 132 animals (Stear et al., 1988b). In one herd of Zebu
cross cattle (247 individuals) both groups of parasites were considered
(Hetzel et al. 1988). The number of alleles analysed in the three studies
were 16 (11), 24 and 17, respectively. In the two latter investigations
the model for the single allele procedure included fixed sire effects. In
conclusion, no strongly significant differences were observed. The analysis of segregation in halfsib families indicated, however, a possible
effect of w16 with respect to reducing the number of eggs and this was
consistent with previous results (Stear et al., 1984). The allele w16
(CA27) seemed also to confer resistance to tick infestation (not significant) (Holroyd and Stear, 1985).

CONCLUSION

Based on the review it may be concluded that there seems to be difficulties in demonstrating significant and/or consistent MHC associations by
means of population data. Several reasons for this may exist and here four
will be mentioned:

1) The genetic basis for studying the association between alleles at
one MHC locus, in the reviewed investigations a Class I locus, and variation in a particular quantitative trait may be that the locus is the only
major locus for the trait, i.e., the disease gene belongs to this locus;
that the locus is one of a few major loci possibly showing some epistatic
interactions; or that the locus does not directly influence the trait considered but is a marker locus for the trait due to gametic disequilibrium
or linkage to a major locus, e.g., a Class II locus. For the latter two
hypotheses associations to different alleles in different populations are
likely and so studies of inheritance are necessary to elucidate the
influence of the considered locus.

2) The same disease complex may be studied by means of several disease
parameters. As an example, subclinical mastitis may be assayed as infected
quarter, as increased SCC or as changes in several other indirect inflammation markers. Basic biological studies are needed to find out which of
these parameters, if any, are most likely to be influenced by the products

of the MHC region. If possible, continuous disease parameters should be chosen since they give the most detailed information about the variation between individuals.

3) The investigations reviewed have used field data including halfsib families except for the case-control study of persistent lymphocytosis. The data were very unbalanced with respect to the distribution of MHC genotypes and alleles on external and genetic factors influencing the variation in the trait so they were suboptimal for drawing inference about the effect of MHC genotypes and alleles. The statistical power of tests for MHC associations may be increased very much by considering infection experiments with groups of animals with only a few different alleles or by other planned experiments with proper randomization. Finally, the single allele procedure seems to be insufficient in demonstrating small effects.

4) If the MHC polymorphism has been build up over many, many generations due to small heterozygous advantages then the differences between allelic effects may in general be very small (Klein and Figueroa, 1986). In addition, cattle populations are under constant selection pressure for disease resistance so susceptibility genes with large effect may be very rare.

Despite these general problems there still may be particular alleles and traits that are strongly associated. Such alleles may be interesting for basic research on disease mechanisms in that they may give information about the etiology of the diseases. In human medicine the MHC alleles are used as diagnostic tools in a few cases. A similar application of MHC associations in cattle is unlikely. A main purpose of investigating MHC associations in farm animals has been to find associations which could be used in breeding for more resistent animals. At this moment this purpose cannot be fulfilled in cattle. In the future, MHC genotyping may be used to control that breeding populations maintain a reasonable high frequency of alleles associated with disease resistance.

ACKNOWLEDGEMENT

The data on somatic cell counts were kindly made available by N.E. Jensen. The work has been supported be The Animal Biotechnology Research Center in Denmark.

REFERENCES

Hetzel, D.J.S., Stear, M.J., Nicholas, F.W., Mackinnon, M.J. and Brown, S.C. 1988. Parasite resistance and the major histocompatibility system in cattle. Animal Genetics, in press.

Holroyd, R.G. and Stear, M.J. 1985. The role of the bovine leukocyte anti-
gen (BoLA) complex in resistance to tick parasites. In: W.C. Davis,
J.N. Shelton and C.W. Weems (Editors), Characterization of the Bovine
Immune System and The Genes Regulating Expression of Immunity with
Particular Reference to Their Role in Disease Resistance. University
of Washington Press, pp 165-172.

Kaartinen, L., Ali-Vehmas, T., Larsen, B., Jensen, N.E. and Sandholm, M.
1988. Relation of the bovine M blood group system with growth of
Streptococcus agalactia and Staphylococcus aureus in whey. J. Vet.
Med. B., 35, 77-83.

Klein, J. and Figueroa, F. 1986. Evolution of the major histocompatibility
complex. CRC Crit. Rev. Immun., 6, 295-386.

Larsen, B., Jensen, N.E., Madsen, P., Nielsen, S.M., Klastrup, O. and
Madsen, P.S. 1985. Association of the M blood group system with bovine
mastitis. Anim. Blood Groups Biochem. Genet., 16, 165-173.

Lewin, H.A. and Bernoco, D. 1986. Evidence for BoLA-linked resistance and
susceptibility to subclinical progression of bovine leukaemia virus
infection. Anim. Genet., 17, 197-207.

Lewin, H.A., Wu, M.-C., Stewart, J.A. and Nolan, T.J. 1988. Association between BoLA and subclinical bovine leukemia virus infection in a
herd of Holstein-Friesian cows. Immunogenetics, 27, 338-344.

Oddgeirsson, O., Simpson, S.P., Morgan, A.L.G., Ross, D.S. and Spooner,
R.L. 1988. Relationship between the bovine major histocompatibility
complex (BoLA), erythrocyte markers and susceptibility to mastitis
in Icelandic cattle. Animal Genetics, 19, 11-16.

Østergård, H., Kristensen, B. and Andersen, S. 1988. Investigations in
farm animals of associations between the MHC system and disease
resistance and fertility. Livest. Prod. Sci., in press.

Solbu, H., Spooner, R.L. and Lie, Ø. 1982. A possible influence of the
bovine major histocompatibility complex (BoLA) on mastitis. In:
Proceedings of 2nd World Congress on Genetics applied to Livestock
Production (Madrid), vol. 7, 368-371.

Spooner, R., Morgan, A., Sales, D., Simpson, P., Solbu, H. and Lie, O.
1988. MHC associations with mastitis. Animal Genetics, 19 suppl. 1,
57-58.

Stear, M.J., Nicholas, F.W., Brown, S.C., Tierney, T. and Rudder, T.
1984. The relationship between the bovine major histocompatibility
system and faecal worm egg counts. In: J.K. Dineen and P.M. Out-
teridge (Editors), Immunogenetic Approaches to the Control of Endopa-
rasites with Particular Reference to Parasites of the Sheep. CSIRO,
pp. 126-136.

Stear, M.J., Dimmock, C.K., Newman, M.J. and Nicholas, F.W. 1988a. BoLA
antigens are associated with increased frequency of persistent
lymphocytosis in bovine leukaemia virus infected cattle and with increased incidence of antibodies to bovine leukaemia virus. Animal
Genetics, 19, 151-158.

Stear, M.J., Tierney, T.J., Baldock, F.C., Brown, S.C., Nicholas, F.W. and
Rudder, T.H. 1988b. Class I antigens of the bovine major histocompatibility system are weakly associated with variation in faecal worm egg
counts in naturally infected cattle. Animal Genetics, 19, 115-122.

POSSIBLE EFFECTS OF THE PIG SLA COMPLEX ON PHYSIOLOGICAL PERFORMANCES

Marcel VAIMAN
Laboratoire de radiobiologie Appliquée - IPSN-DPS-SPE
91191 GIF SUR YVETTE - Cedex - FRANCE

ABSTRACT.

Associations have been found between the swine major histocompatibility complex SLA on the one hand and carcass composition and growth performances on the other. The SLA complex appears to influence a variety of reproductive parameters, including the ovulation rate, preimplantation embryonic development and litter size. Piglet viability and weight at birth and weaning were also found to be associated with certain SLA haplotypes. Prolificacy was affected in matings in which boars and sows had common SLA haplotypes, and in males, the SLA was found to affect sex organ development.

INTRODUCTION

The two major families of genes of the major histocompatibility complex (MHC) encode cell-membrane polymorphic glycoproteins whose role is to activate the immune system machinery by presenting processed antigens to T cells. Besides having a predominant role in immune physiology, the MHC region has been shown to take part in various non immune biological events such as production and hatching in the chicken (Simonsen et al. 1982 ; Morton et al. 1965). As regards mammals, the influence of the HLA complex in sterility is a controversial issue in man, but the effect of the RT1 linkage group on fertility and embryonic development in the rat is well documented (Kuntz et al. 1980). Similarly, in the mouse, traits related to reproduction have been shown to be controlled by genes of the H-2 complex. For instance ,an H-2 linked gene was called PED, in reference to its role in the development of preimplanted mouse embryos (Warner, 1986). Thus the MHC appears to affect productive and reproductive traits in laboratory and farm species, but neither the genes nor the mechanisms responsible have yet been identified.

These observations prompted us to investigate the role of the MHC in zootechnical performances in the pig. In addition, markers for productive and more importantly for reproductive traits are needed in order to improve swine prolificacy.

The genetic organization of the SLA complex is beginning to be fairly well understood. As in other mammals, the SLA complex in the pig comprises genes coding for molecules of the three MHC classes (Chardon et al, 1988). However, the swine SLA complex displays an unusual feature, as the genome only contains 7 or 8 DNA class I sequences. The number of class II sequences appears to be comparable in pig and other mammals. As far as we know, there seems to be a single C2 and Bf gene (Lie et al., 1987) as well as one C4 and one 21-hydroxylase gene in the SLA complex.

SLA class I serology is now well developed and allows characterization of most of the existing haplotypes in commercial breeds. A SLA haplotype chart was recently etablished as a result of the cooperation of several European laboratories (Renard et al. 1988). However as only the class I specificities are readily definable, comparisons between unrelated haplotypes must be made with caution.

SLA COMPLEX AND IMMUNE RESPONSES

As expected the SLA complex was found to constitute the strongest transplantation barrier, while the A and E blood group systems behaved like minor histocompatibility systems.

The SLA complex has been shown to influence the immune response to various antigenic compounds including complex antigens like bordetella bronchiseptica, and also to interfere in complement hemolytic activity (Vaiman et al. 1988). Similarly there is good evidence that the SLA complex influences the expression of an inheritable swine cutaneous malignant melanoma (Tissot et al. 1987).

SLA AND PRODUCTION TRAITS.

Several years ago we reported preliminary results which suggested that the SLA region might influence growth rates and carcass composition. A more recent investigation in groups of unrelated French Large White and Landrace young gilts provided new evidence that the SLA region influences body composition, especially lipid metabolism as shown in table 1.

Table 1 : Significant associations between SLA haplotypes and carcass characteristics.

Breeds and number	Large White (139)	Landrace (100)	
Haplotypes and number of swine with haplotype	H04 (31)	H04 (23)	H23 (23)
carcass length (mm)		-12.79^{*} +/-5.99	
shoulder (kg)	$+0.180^{***}$ +/-0.055	-0.184^{**} +/-0.074	
ham (kg)			$+0.179^{*}$ +/-0.090
loin (kg)	$+0.301^{+}$ +/-0.162		$+0.250^{*}$ +/-0.156
belly (kg)			-0.149^{*} +/-0.081
Backfat (kg)	-0.542^{***} +/-0.158	$+0.310^{*}$ +/-0.143	-0.471^{***} +/-0.145
Fat thickness			
rump (mm)	-3.08^{***} +/-0.94		
back (mm)	-2.23^{**} +/-0.76		-1.39^{*} +/-0.71
shoulder (mm)	-3.47^{***} +/-1.03	$+2.18^{*}$ +/-1.04	-3.31^{***} +/-1.06
Fat per cent	-2.49^{**} +/-0.86	$+1.36^{*}$ +/-0.76	-2.47^{***} +/-0.76
lean per cent	$+1.86^{**}$ +/-0.72	-1.02^{+} +/-0.64	$+2.11^{***}$ +/-0.64
loin/backfat rate	$+0.488^{**}$ +/-0.183	-0.26^{*} +/-0.13	$+0.384^{**}$ +/-0.136

As shown in table 1, the carcasses of the Large White females bearing haplotype H-04 and of the Landrace breed bearing haplotype H-23 had a significantly lower - than - average fat content. In addition these animals had a meat content 2 per cent higher than that of those bearing haplotypes other than H-04 in Large White and H-23 in Landrace. In the latter breed, however, haplotype H-04 was associated with greater carcass fatness. In addition, significant associations have been found in Large White females between some SLA haplotypes and features reflecting meat quality, especially reflectance and H-16, H-8 and H-4 haplotypes. Similarly, haplotype H-02 was associated with score and water holding capacity .

SLA AND REPRODUCTIVE TRAITS

The first indication that the SLA region might influence pig reproduction was the observation of a deficit of SLA homozygotes in litters,
suggesting some kind of selection favoring heterozygotes (Vaiman and
Renard, 1980). Since this original finding, many investigations have
been carried out in different pig populations on the presumptive role of
the SLA complex in reproduction and foetus and piglet development. As
shown in table 2, a wide range of parameters was assessed.

Table 2 : List of association found between the SLA complex and
pathological and physiological reproductive traits.

1-Ovulation rate	7-Piglet weaning weight
2-Preimplantation embryonic development	8-Piglet survival rate
3-Litter size at birth	9-Prolificacy
4-Homozygotes at birth.	10-Male sexual tract development
5-Litter size at weaning	11-Segregation distorsion
6-Piglet birth weight	12-Tissue androstenone level

Among the variables studied the ovulation rate seems to be significantly affected. Rothschild and colleagues found that in an experimental
herd of commercial pigs selected over many generations for its ovulation
rate, a significant shift occurred in the frequency of SLA haplotypes,
compared to the control line (Rothschild et al. 1984). The same authors
subsequently adopted a more direct approach by counting the corpora
lutea in miniature swine dams which had been inbred for three distinct
SLA haplotypes (Conley et al. 1988). They found that the ovulation rate
was higher in sows homozygous for the d haplotype than in sows with the
SLA a/a and a:c genotypes ($P < 0.01$). In addition, the SLA d/d females
tended to have more embryos although the numbers recovered between days
2 and 6 after mating did not differ significantly among the three SLA
dam genotypes.

It is interesting to note that the litters from sows bearing one or
two d haplotypes or sows mated to a SLA d/d boar were significantly
larger than those of other matings ($P < 0.05$). On the other hand, SLA
genotypes had no noticeable effects on the concentrations of
oestradiol-17 beta and oestrone in peripheral and uterine blood in the
three lines of miniature pigs.

Reports dealing with the associations between SLA and litter size in commercial breeds led to somewhat contradictory conclusions (Vaiman et al. 1988). The overall impression was that litter size at birth and weaning was at most only slightly influenced by the SLA region, with possible exceptions. Thus sows homozygous for haplotype SLA H-16 tended to have small litters, whereas females carrying the haplotype H-6 were found to have consistently larger litters than SLA H-6 negative sows. Further, the frequency of the SLA H-6 haplotype among a group of French hyperprolific sows, selected on the basis of three successive farrowings with at least seventeen piglets born, each time, was about three times greater than in the overall population (15 v.s. 5 per cent). However, since the SLA H-6 haplotype appears to influence growth performances in males negatively, counter selection against it may have contributed to its disappearance from one of the herds investigated.

Although no large scale studies have yet been made, several findings appear to confirm that SLA homozygosity in piglets may affect litter size negatively.

This may be connected with investigations showing that matings between boars and sows with identical SLA haplotypes occasionally gave smaller litters than matings between fully SLA disparate parents. The reasons are not yet known. From preliminary results for miniature pigs it can be inferred that the SLA region may influence the embryonic development rate (Schwartz et al. 1988). Thus, on day 6 of gestation, embryos from SLA d dams contained fewer blastomeres, (23 cells on an average), than embryos from SLA a or SLA c sows (89 and 79 cells respectively). Fewer blastomeres were also noted in SLA d-dams than in dams with the two other haplotypes at days 9 and 11. Since the embryos that develop fastest appear to have a selective advantage over the others, the embryo development rate might be an important variable.

That this rate may be affected by the SLA complex has been documented in a Large White breed. In one study involving the SLA H-2 haplotype, litter size reduction was combined with a significant reduction in the number of SLA H-2 homozygous piglets at birth compared to the number expected, as shown in table 3.

Table 3 : Influence of the SLA H2 Haplotype on embryonic
development and litter size.

Mating type	Embryonic development at 31 &32 days			
	Placenta weight(g)	Amniotic fluid(ml)	Embryo weight(g)	Embryo length(mm)
boars sows H2/- x H2/-	27.8*(70)	171*	1.76*	25.9*
H2/- x -/-	43.5 (59)	208	2.69	29.2

	30-35 days embryo loss (%)	Litter size at birth (n)	H2/H2 piglets at birth (%)
H2/- x H2/-	35.5 * (17)	9.86* (6)	9*
H2/- x -/-	19.8 (27)	12.75 (12)	

* P < 0.05 . The SLA H2 haplotypes found in the boars and sows were of the same origin. a : number of embryos.b: embryos found at 30-35 days compared to the number of corpora lutea.c: number of litters.d:instead of 25% expected.

A gamete selection mechanism during fecundation and/or a higher mortality rate in pregnancies involving SLA H-2 homozygous embryos might account for this observation. To clarify this question, planned matings were carried out in order to investigate the development rate and the embryonic loss before day 35 of gestation in groups with or without the expected SLA H-2 homozygous embryos. Table 3 shows that a embryonic development dropped significantly in the group with SLA 2 homozygous embryos from the 31st day of gestation onwards. More important was the finding that the percentage of embryo loss in this group at day 35 was twice that observed in the group without SLA H-2 homozygous embryos. Before day 30, however, there was no significant difference between the two groups. A simple calculation shows that the higher mortality rate found in the group with SLA H-2 homozygous foetuses may alone account for the litter size reduction noted at birth. Therefore, it appears that no further selective losses occurred after day 35 of gestation. Although the precise mechanism responsible for the greater losses in the SLA H-2 homozygous group is not known, the existence of recessive lethal genes within the SLA region can be suggested.

The SLA complex appears to interfere significantly during both foetal life and post-natal development. Negative or positive effects have been observed, depending on the haplotypes considered. Thus, Landrace sows bearing the haplotypes H-23 had piglets which were lighter than average at birth and to some extent at weaning. It is not yet clear whether this effect is related to the fact that, as mentioned above, the haplotype H-23 has been found to be significantly associated with reduced carcass fatness in Landrace females. On the other hand, piglets bearing the haplotype H-12 were found to be heavier than average at birth and even more so at weaning.

No correlations could be established either between the SLA markers and mating failures ,or between the SLA complex and still-birth rate. On the contrary the mortality rate before weaning was lower in piglets bearing the haplotype SLA H-1, although it rose significantly after weaning. The increased mortality in SLA H-1 piglets appeared to be connected with a higher rate of fatal diarrhea, whose ethiology was, however, not established.

SLA segregation distorsion has been reported in Landrace animals.

The influence of the SLA region on male sexual tract development was unexpected. The males whose SLA phenotypes were characterized were primarily selected for their high or low tissue content of androstenone, a sexual steroid compound which is a pheromone in swine and is responsible for boar taint. Although the SLA complex did not markedly affect the androstenone level, two haplotypes-SLA H-15 and SLA H-16-had significant positive effects on testes, epididymes and Cowper's gland development. On the other hand the SLA H-4 haplotype was found to affect these organs negatively. Thus, the SLA region is in some way related to the metabolism of male sex hormone target tissues. It is of interest to note that testes weight in boars has been found to correlate with the ovulation rate of related sows.

DISCUSSION

The data accumulated during the last decade are strongly suggestive of the influence of the SLA region upon a large number of traits, some of economic importance. These findings are encouraging for future research, as better knowledge of the swine MHC may eventually provide a means for more efficient selection programmes designed to improve zootechnical performances.

The mechanisms responsible for the involvement of the SLA region in productive and reproductive traits are not yet known. It is not clear how class I and class II products could influence non-immune cell physiology. Nevertheless, there have been concordant reports on physiological associations between human class I MHC antigens and several cell surface receptors, including epidermal growth factor and insulin receptors.

Such observations provide clues as to how the MHC antigens might be involved more generally in cellular and tissue physiology.

SLA complex genes other than those of class I and class II can be considered as particularly relevant to the variety of effects observed. Thus, the gene coding for the 21-hydroxylase enzyme or cytochrome P450, one of the main enzymes involved in steroid biosynthesis, would be particularly relevant in this respect.

Another important family of genes was recently mapped to the HLA and H-2 class III region. These genes code for two related polypeptides, the tumor necrosis factor alpha and beta molecules, which proved to be cytokines with pleiotropic activity (Carroll et al. 1987 ; Muller et al. 1987).

Among many other functions, these molecules act as growth factors and are even involved in the physiological regression of the corpus luteum. Whether TNF-like genes also map inside the SLA complex has not yet been established, but there is good reason to believe that this is so, on the basis of to what is known about humans and mice. An additional gene was very recently discovered thin the class III region of HLA and H-2 (Levi-Strauss et al. 1988). It apparently codes for an unusual protein, whose role is still unknown, but illustrates once more the extreme complexity of the MHC region, as well as the great degree to which this region is preserved during evolution.

Finally, the role of the SLA complex in specific and non specific immune defences should not be overlooked.

REFERENCES.

Carroll, M.C., Katzman, P., Alicot, E.M., Koller, B.H., Geraghty, D.E., Orr, H.T., Strominger, J.L., and Spies, T. 1987. Linkage map of the human major histocompatibility complex including the tumor necrosis factor genes.Proc. Natl.Acad.Sci. USA. 84,8535-8539.

Chardon, P., Geffrotin, Cl., Vaiman, M. 1987. Genetic organization of the SLA complex. Proceeding of the International Conference on the Major Histocompatibility Complex of Domestic animal species. Iowa State Press, in press.

Conley, A.J., Jung, Y.C., Schwartz, N.K., Warner, C.M., Rothschild M.F. an Ford, S.P. 1988. Influence of SLA haplotype on ovulation rate and litter size in miniature pigs. J. Reprod. Fert. 82, 595-601.

Kunz, H.W., Gill,T.J., Dixon, B.D., Taylor, F.H., and Greiner, D.L. 1980. Growth and reproduction complex in the rat. Genes linked to the major histocompatibility complex that affect developent. J; Exper. Med. 152, 1506-1518.

Schematic representation of the genes of the Major Histocompatibility Complex (CMH), their products on cellular membranes (Macrophages) and effects on cells of the immune system (T and B cells) and their involvement with corticosteroid production (original by M. Vaiman).

IL	–	interleukin	C	– cortex
IFN	–	interferon	CE	– capilliary endothetial cells
TNF	–	tumor necrosis factor	M	– medulla or macrophage
CRF	–	corticotropin releasing factor	C 1,2,3,4	– complement components
ACTH	–	adrenocorticotropin	B	– antibody producing lymphocytes
BCG	–	bacille Calmette-Guérin	T	– lymphocytes for T-cell mediated immunity
LPS	–	lipopolysaccharide		

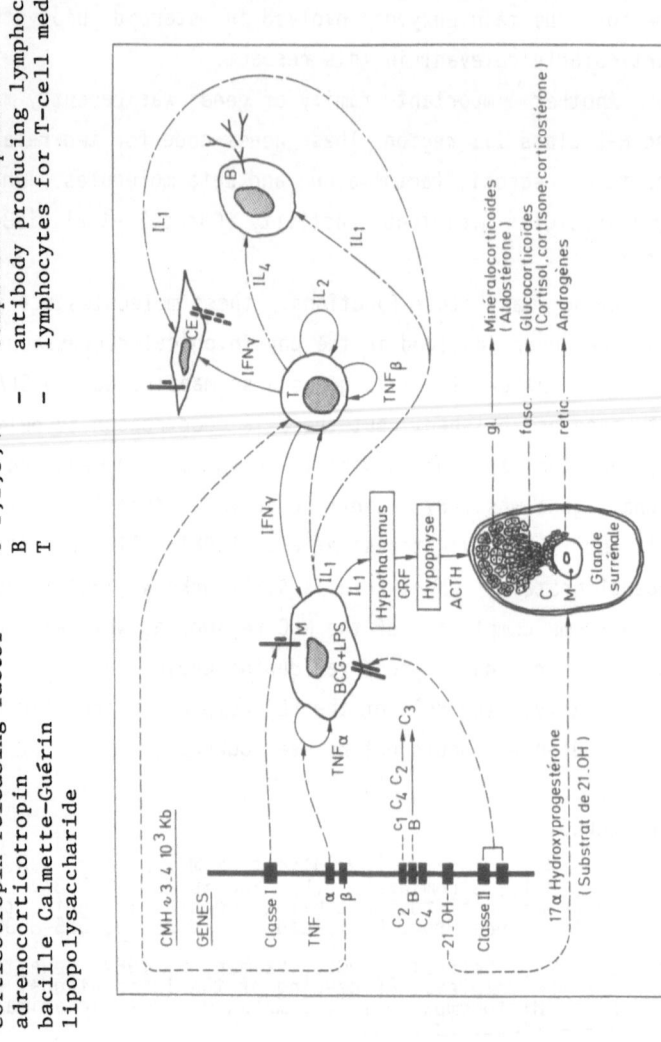

Lévi-Strauss, M., Carroll, C., Steinmetz, M., and Meo, T. 1988. A previously undetected MHC gene with an unusual periodic structure. Science, 240, 201-204.

Lie, W.R., Rothschild, M.F., and Warner, C,M. 1987. Mapping of C2, Bf and C4 genes to the swine major histocompatibility complex (swine leukocyte antigen). J. Immunol. 139, 3388-3395.

Morton, J.R., GILMOUR, D.G., Mc Dermid, E.M., Ogden, A.L. 1965. Association of blood-groups and protein polymorphisms with embryonic mortality in chicken. Genetics 51, 97-107.

Muller, U., Jongeneel, C.V., Nedospasov, S.A., Lindahl, K.F. & Steinmetz, M. 1987. Tumour necrosis factor and lymphotoxin genes map close to H-2D in the mouse major histocompatibility complex. Nature 325, 265-267.

Renard, Ch., Kristensen,B.,Gautschi, C., Hruban,V. ,Fredholm,M.,Vaiman,M. 1988. Joint report of the first international comparison test on swine lymphocyte alloantigens(SLA). Animal Genetics19,63-72.

Rothschild, M.F., Zimmerman, D.R., Johnson, R.K., Venier, L. and Warner, C.M. 1984. SLA haplotype differences in lines of pigs which differ in ovulation rate. Anim. Blood. Grps. Biochem. Genet. 15, 155-158.

Simonsen, M., Kolstad, N., Edfors-Lilja, I., Liljedahl, L.E. and Sorenson, P. 1982. Major histocompatibility genes in egg-laying hens. Amer. J. Reprod. Immun. 2, 148-152.

Schwartz, N.K., Conley, A.J., Rothschild, M.F., Warner, C.M., Ford, S.P. 1987. Effect of SLA haplotype on rate of preimplantation embryonic development in miniature swine. Abstract n°38. Proc. Intern. Conf on the major histocompatibility Complex of domestic animal species. Iowa State press. in press.

Tissot R.G., Beattie, C.W., and Amoss, M.S. 1987. Inheritance of sinclair swine cutaneous malignant melanoma. Cancer Research 47, 5542-5545.

Vaiman, M. and Renard, Ch. 1980. Deficit of piglets homozygous for the SLA histocompatibility complex in families. Anim. Blood Grps Biochem. Genet. 11 (suppl. 1), 57 (Abstr.)

Vaiman, M., Renard, Ch., Bourgeaux, N. 1987. SLA The major histocompatibility complex in swine : its influence on physiological and pathological traits. Proceeding of the International Conference on the Major Histocompatibility Complex of Domestic animal species. Iowa State Press, in press.

Warner, C.M. 1986. Genetic manipulation of the major histocompatibility complex. J. Anim. Sci. 63, 279-287.

ELA DISEASE ASSOCIATIONS

S. Lazary* and H. Gerber**
*Division of Immunogenetics, Institute for Animal Breeding
and ** Klinik für Nutztiere und Pferde, University of Berne,
3012 Berne, Switzerland

A better definition of the ELA system has made it possible to
carry out association studies for equine diseases. Incomplete
definition of the MHC system in this species and, in allergic
diseases, the lack of a precise etiological diagnosis
(unknown allergens responsible for the same clinical
diagnosis) could be responsible for finding only weak or
negative associations until now. Indeed, susceptibility to
sarcoid is the only disease in horses which to date, shows a
strong association with the ELA system.

INTRODUCTION

Several pathological conditions show associations with
various gene products of the MHC in humans (Tiwari and
Terasaki, 1985). However, a disease association, regardless
of its strength, is a statistical phenomenon which makes no
statement regarding a cause-and-effect relationship between
the allelic MHC gene product and the disease. In most cases
of MHC associated diseases in humans the etiological
factor(s) is unknown. The conditions are characterized by
various malfunctions of the immune apparatus; chronic
inflammatory processes of "autoimmune" nature, involving
humoral and cellular components are the main pathological
findings. One of the basic questions remains unsolved; is
the immunological malfunction in these patients (carrying
certain MHC haplotype) the consequence of a specific effect
of an unknown etiological factor (virus, toxin etc.) on the
immune gene(s) or do these individuals already possess a
defective immunological function (genetically determined by
the MHC) before the appearance of the clinical disorder(s)
leading to the disease in time without the influence of any
external factor.

For MHC-influenced diseases, the finding of associations
has opened new possibilities for genetic studies, at the

134

level of gene products and genes. A better definition of the genetic basis in combination with functional studies could lead to more effective curative and preventive procedures for controlling these diseases.

Our efforts at characterisation of the equine MHC have had the aim, in a later phase, to study the possible influence of this genetic region on some of the main equine diseases. Diseases with autoimmune components or of auto-immune nature (Pemphigus foliaceus) appear to be rare in the horse. On the other hand, chronic inflammatory disorders of the joints with unknown etiology are interesting candidates for future studies in combination with the ELA system. Disorders of allergic nature exist in horses: allergic pneumopathies and summer eczema. Allergic conditions in humans do not show strong associations with the MHC-region. The results of our investigations in horse are summarized in the present paper. Up to now, equine sarcoid, a skin tumor, shows the strongest association with the ELA system in horse. In this paper the latest results of the association between ELA and sarcoid susceptibility is presented.

MATERIAL AND METHODS

The animals and methods used are given in the cited original papers. In the present paper the demonstrated association between ELA and sarcoid susceptibility is based on 224 Swiss, French and Irish warmblooded Hunter-type horses with clinically and partially histologically diagnosed sarcoid. The 609 control horses had the same age and sex distribution. They represented the same breeds as the groups of affected horses, but there were no detectable sarcoids, nor was there a history of previous sarcoid affection.

ELA specificity determination was performed by the microcytotoxicity test according to Bernoco et al.(1987). The typing results were computed according to Svejgaard et al.(1983) for each ELA specificity and breed for breed. Relative risk and etiological fraction were calculated.

For statistical significance, the Chi square test was applied.

In multiple-case families, at least three members (parent(s), full- and/or half-siblings etc.) were affected by sarcoid. In most cases the families consisted of half-siblings. In these cases only the two ELA alleles (or haplotypes) of the common ancestor in question were analyzed for an association with disease susceptibility. Since the antigens A5, W20 and W13 occurred with increased frequencies in affected animals, suggesting a contribution to the pathogenesis of the disease, offspring which inherited one of these antigens (haplotypes) from their other ancestor were excluded from the statistical analysis as non-informative. Apparent homozygotes (the other ancestor could not be typed) were included in our calculations.

The Chi square test was carried out separately on affected and healthy informative offspring on the basis of the number of observed/expected haplotypes transmitted. Offspring from families with common ELA antigens were combined for calculation. In the Chi square analysis performed for the control population the distribution of ELA phenotypes demonstrated a good fit to Hardy-Weinberg expectation: $P > 0.4$.

RESULTS

ELA/Laminitis

Two studies on the distribution of ELA antigens in animals suffering from laminitis have been published (Lazary et al., 1985a; Meredith et al., 1986). The same frequencies of ELA antigens were observed in both diseased and control groups.

ELA/Allergic Pneumopathy

In the single study to date (Lazary et al., 1985a), diseased animals (N:162) displayed the same distribution of the ELA antigens as the control group. The occurrence of this disease seems to be more frequent in some families than in others.

ELA/Summer Dermatitis

Affected groups of Iceland ponies (N:116) living in different European countries showed, in comparison to controls (N:178), an increased frequency of the second ELA locus coded W22 specificity: RR: 2.04, X^2: 5.88 (Haldersdottir, pers. communication).

Accumulation of cases in certain Haflinger- and Arabic horse families and breeding lines was observed but typing result suggest, that besides the ELA system other genetic factor(s) also influence the manifestation of this disease (unpublished results).

ELA/Sarcoid

To date, sarcoid disease has shown the strongest association with the MHC region in the horse. Studies have been performed at the population level (Lazary et al., 1985b; Meredith et al., 1986, Broström et al., 1988) in Swiss, French, Irish and Swedish warmblooded horses and in Thorough-breds. ELA W13, a second locus specificity shows the strongest association throughout these breeds (results given in Table 1). Further analysis of the typing results of the diseased groups showed clearly that diseased animals not carrying the W13 antigen displayed however, one or the other (A5 or W20) ELA antigen at higher frequency than the controls, instead of W13.

To clarify the importance of the other ELA haplotypes (without W13) for sarcoid susceptibility, typing studies with sarcoid diseased and healthy offspring of W13 carrying and non-carrying descendents (Gerber et al., 1988; Broström et al. 1988) were commenced. The results show clearly that in families with multiple sarcoid cases the predisposition is associated with certain haplotypes within families. However, the disease associated ELA haplotype can contain various ELA antigens (A3 with W13, A5, A15 with or without W13, W20 or A1

TABLE 1 Distribution of sarcoid-associated equine MHC antigens in Swiss, French and Irish warmblooded horses. ELA-A3 and -A5 are first locus alleles (19 alleles are presently defined). ELA-W13 is a second locus allele (4 alleles presently defined).

Breed	Affected (N)	Controls (N)	ELA	Antigen frequency(%) affected	controls	RR	EF	X^2	cP
Swiss	158	361	A3	39	30	1.45	.12	3.53	n.s.
			A5	32	22	1.63	.12	5.26	n.s.
			W13	45	25	2.46	.27	20.56	<0.001
French	27	120	A3	63	29	4.13	.48	11.01	<0.05
			A5	33	22	1.81	.15	1.65	n.s.
			W13	78	28	9.28	.69	23.97	<0.001
Irish	39	128	A3	59	43	1.91	-.28	3.08	n.s.
			A5	23	26	0.86	-.04	0.12	n.s.
			W13	62	34	3.16	.42	9.72	<0.05

RR: Relative risk; EF: Etiologic fraction; cP: corrected P

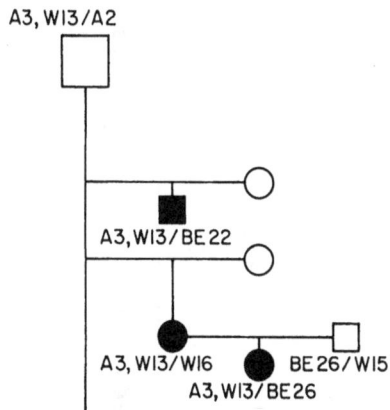

Fig. 1   Typing results of affected offspring of
stallion No. 1. Symbols: circles, females; squares,
males; black symbol indicates that the individual is
affected with sarcoid.

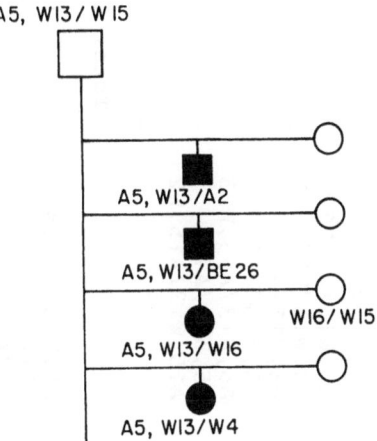

Fig. 2.   Typing results of affected offspring of
stallion No. 3.

TABLE 2 Distribution of the transmitted ELA haplotypes in informative sarcoid-affected/healthy offspring and Chi square test for distorsion of haplotype segregation.

Parents	Parental haplotypes 1. 2.	sarcoid/healthy	Sarcoid with parental haplotype 1.	2.	Healthy with parental haplotype 1.	2.
Stallion 1	A3,W13/A2	3/6	3	0	4	2
Stallion 2	A3,W13/A3,W12	2/22	2	0	10	12
Stallion 9	A3,W13/W20	4/26	4	0	12	14
Stallion 10	A3,W13/A1	3/2	3	0	1	1
Stallion 11	A3,W13/W18	2/0	2	0	-	-
Stallion 6	A5,W13/A15,W23	4/23	3	1	10	13
Stallion 3	A5,W13/A15	4/3	4	0	1	2
Stallion 12	A15,W13/A15,W23	5/3	4	1	2	1
		Total 27/85	25	2	38	45
				P<0.001		45 n.s.
Stallion 4	A5,W23/A19,W23	11/4	8	3	8	6
Mare 1	A5/A6	2/4	2	0	2	2
Stallion 7*	A5/A2,W22	5/4	4	1	2	2
Stallion 8	A5/A2	3/4	3	0	2	2
		Total 21/16	17	4	14	12
				P<0.05		12 n.s.

* Thoroughbred family

in Arabs without W13), varying from family to family. The typing results for offspring of two stallions is shown in Figures 1 and 2. From the heterozygous sires, the A3,W13 and the A5,W13 haplotypes were transmitted to the offspring together with the sarcoid susceptibility. A summary of the typing results of families with W13 and with A5,W23 haplotypes is shown in Table 2.

DISCUSSION

Laminitis and allergic bronchitis do not seem to be associated with the ELA system. For both these disorders on the other hand, the etiological factors are not well defined for the individual diseased animals and the groups are classified solely on the basis of clinical symptoms . The same classification problem is also true for the summer dermatitis affected animals; different allergens from Culicoides, simulids etc. could be the etiological factors in any individual case. In all these studies with allergic diseases we need a better grouping on the basis of the actual specific allergens.

Up to date, the susceptibility to equine sarcoid is the only disease which in different breeds and countries (USA, Sweden, Switzerland) shows the same strong association with the MHC region at the population level and in family studies.

As the family data demonstrate, there is not just one class I or II ELA antigen involved, but possibly an unknown gene(s) linked to this region. Another possibility is that a "Sarkoid susceptibility gene(s)" does not exsist as such, but rather that the development of this disease is based on the specific immunological response of the host against the viral etiological factor, which on the other hand is specifically influenced by the different haplotypes.

ACKNOWLEDGMENT
This work was supported by the Swiss Nationalfonds for Scientific Research, grant 3.879-0.88

REFERENCES

Bernoco, D., Antczak, D.F., Bailey, E., Bell, T.K., Bull,
    R.W., Byrns, G., Guerin, G., Lazary, S., McClure, J.,
    Templeton, J. and Varewyck, H. 1987. Joint report of the
    fourth international workshop on lymphocyte alloantigens
    of the horse. Held in Lexington, Kentucky 19-20 October
    1985. Animal Genetics 18:81-94.
Broström, H., Dubath, M.L., Lazary, S., Larsson, A. and
    Perlmann, P. 1988. Equine leucocyte antigens and
    epidemiology of equine sarcoid. Veterinary Immunology
    and Immunopathology (submitted).
Gerber, H., Dubath, M.L. and Lazary, S. 1988. Equine
    Infectious Diseases V. In Press.
Lazary, S., Glatt, P.A. and Gerber, H. 1985a. Leucocyte
    antigens in various pathological conditions in horses.
    Animal Blood Groups and Biochemical Genetics 16:
    Supplement 1 92.
Lazary, S., Gerber, H., Glatt, P.A. and Straub, R. 1985b.
    Equine leucocyte antigens in sarcoid-affected horses.
    Equine vet. J. 17:283-286.
Meredith, D., Elser, H.A., Wolf, B., Soma, R.L., Donawick,
    J.W. and Lazary, S. 1986. Equine Leukocyte Antigens:
    Relationships with Sarcoid Tumors and Laminitis in Two
    Pure Breeds. Immunogenetics 23:221-225.
Svejgaard, A., Platz, P. and Ryder, L.P. 1983. HLA and
    disease 1982. A Survey. Immunological Rev. 70:193-218.
Tiwari, L.J. and Terasaki, I.P. 1985. HLA and Disease
    associations. Springer Verlag New York, Inc.

SESSION 5

Immune response markers and disease resistance

Chairperson: Dr. E. Andresen

AN IMMUNE COMPETENCE PROFILE IN SWINE

H. Buschmann and J. Meyer

Institut für Med. Mikrobiologie, Infektions-
und Seuchenmedizin und Institut f. Tierzucht
und Tierhygiene der Tierärztlichen Fakultät
der Universität München

ABSTRACT

An immunocompetence profile has been developed for pigs. In vitro and in vivo immunoassays were used to determine several immunological parameters being assumed to have direct impact on disease resistance. The repeatabilities of the parameters tested were estimated and the correlations among the single traits were calculated.

INTRODUCTION

There is a great variation in the immune responsiveness of normal pigs both after artificial immunization and after natural infection. A considerable part of this variation is genetically determined. It would be of great practical implication to use this genetic variation for breeding animals of superior resistance to infectious diseases. The critical questions still open are: Which immunological parameters can be successfully used as markers of increased disease resistance ? What is the environmental and genetic variance of these parameters ?

Though our knowledge about the involvement of the different compartments of the immune system in resistance to infectious diseases is very limited, it is well established that T lymphocyte dysfunction is the cause of severe infections with viral pathogens whilst B lymphocyte dysfunction results in infections with the more common bacterial pathogens; neutropenia or deficiency of granulocyte function results in repeated bacterial infections.

We made an approach to define an immunocompetence profile in pigs by measuring several parameters of cellular and humoral immunity in blood samples drawn from living animals.

MATERIAL AND METHODS

Pigs from the breeds German Edelschwein (DE), Pietrain (Pi), German Landrace (DL) and crosses among these breeds were tested at an age of 8 - 12 weeks. All animals were kept under comparable environmental

145

conditions on the same farm. From each animal 20 ml blood samples were
drawn, heparin was used as an anticoagulant.

The aim of the laboratory tests was:

- the determination of the cellular constituents of the blood, i.e. total
  number of leucocytes, differential blood cell counts, percentage of
  rosetteforming cells, sIg$^+$ cells, characterization of the lymphocyte
  subpopulations by monoclonal antibodies;
- the evaluation of functional tests measuring immunocompetence, i.e.
  mitogenic stimulation of lymphocytes, various phagocytosis assays,
  natural killer cell activity (NK)against K 562 target cells;
- determination of Ig isotype concentration in blood plasma by laser
  nephelometry.

The techniques for the determination of the cellular blood constituents and
of the serum Ig values were described previously (Buschmann et al.,1985).
The mitogenic stimulation of lymphocytes was measured in a ^{14}C-thymidine
incorporation assay. The following mitogens were used: Phytohemagglutinine
(PHA) 20 ug/ml; Concanavalin A (Con A) 100 ug/ml; Pokeweed Mitogen (PWM)
100 ug/ml; Bacterial Lipopolysaccharide from Salmonella typhosa (LPS)
2 mg/ml. Always 100 ul lymphocyte suspension (5×10^6 / ml) were incubated
with 100 ul mitogen solution (controls containing only medium).

Concerning phagocytosis assays the following yeast cell killing
system was newly established: 100 ul Baker's yeast suspension (10^6/ml)
were incubated 1^h (37° , continuous agitation) with 100 ul autologous
plasma, 100 ul HBSS and 100 ul granulocyte suspension (7×10^6 / ml).
Then 100 ul 2.5% cold sodium desoxycholate solution (pH 8.7) were added
for lysis of the granulocytes. To remove the disturbing DNA released from
the lysed granulocytes 1 ml of a warm (37°) DNAse solution (1 mg / 100 ml
PBS) were added and incubated for 5 min under periodic agitation. After
centrifugation and resuspension of the cell pellet in PBS (w/o Mg^{++} Ca^{++})
30 ul propidium iodide (o.o5 mg / ml PBS) were added and the percentage
of yeast cells with red fluorescence (killed cells) was evaluated in a
laser flow cytometer (Spectrum III from Ortho).

The determination of chemiluminescence was done in an assay using
phorbolmyristate-acetate (PMA) as a soluble initiator of the metabolic
burst in granulocytes (Winter and Buschmann, 1987).

Nitroblue tetrazolium reduction (NBT) was measured in tubes into
which granulocytes (7×10^6 / ml), opsonized zymosan and NBT solution
had been pipetted. After 10 min incubation the reaction was stopped by

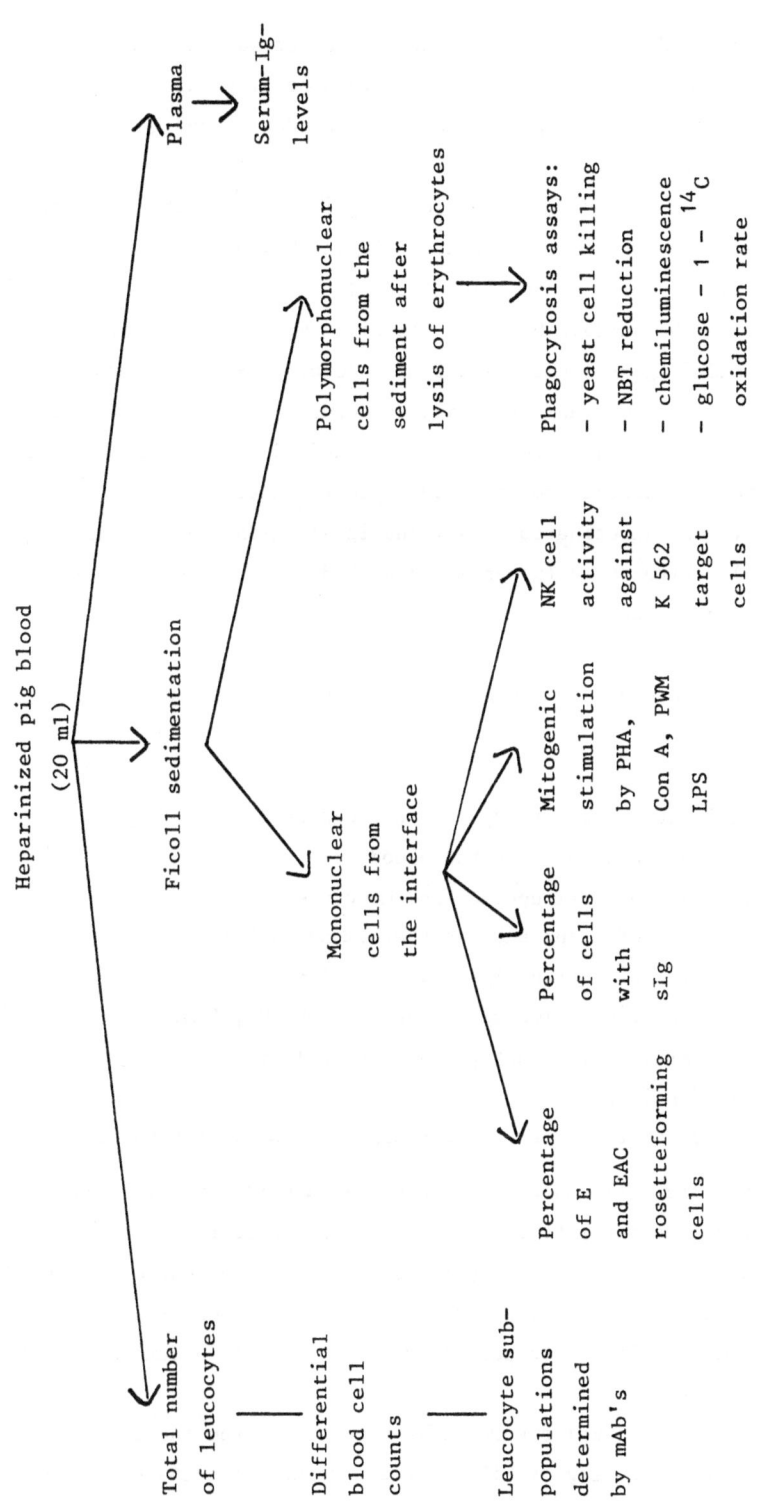

Heparinized pig blood
(20 ml)

Plasma → Serum-Ig-levels

Total number of leucocytes

Differential blood cell counts

Leucocyte sub-populations determined by mAb's

Ficoll sedimentation

Mononuclear cells from the interface

Percentage of E and EAC rosetteforming cells

Percentage of cells with sIg

Mitogenic stimulation by PHA, Con A, PWM LPS

NK cell activity against K 562 target cells

Polymorphonuclear cells from the sediment after lysis of erythrocytes

Phagocytosis assays:
- yeast cell killing
- NBT reduction
- chemiluminescence
- glucose - 1 - ^{14}C oxidation rate

adding cold 1mM N-ethylmaleimide, and the pellet was suspended in N-N-di-methylformamide. Formazan was extracted in a boiling water bath, clarified by centrifugation and the optical density was determined (540 nm) in a spectrophotometer.

The NK-cell activity was measured against Cr-51 labelled K 562 target cells after elimination of adherent cells on FKS coated plastic dishes, the effector : target cell ratio being 100 : 1.

Part of the animals had been immunized with 1 ml tetanustoxoid (150 E / pig) and 5 mg ovalbumine + 2.5 ml incomplete Freund adjuvant (IFA) intramuscularly. Reimmunization was performed 10 days after the first immunization. Antibody titres were determined 10 days after the first and 4 days after the second immunization. Specific antibody activity was tested in a passive hemagglutination assay and in an ELISA assay (Buschmann et al., 1985). With these two antigens we had observed a dose-response relation in pigs and a significant booster effect after repeated injections was seen; further a switch from IgM to IgG antibodies occured in the course of the immune response in the majority of the animals immunized.

RESULTS

Significant breed differences were observed in:
- the number of leucocytes in the blood;
- the percentage of eosinophiles and monocytes;
- the percentage of E and EAC rosetteforming cells;
- the percentage of surface-Ig positive cells;
- the mitogenic stimulation of lymphocytes by PHA, PWM, LPS;
- phagocytic killing of yeast cells by granulocytes;
- the Ig levels in serum;
- the antibody titres after primary immunization with tetanus toxoid.

Systematic influences of sex and season were obvious. Only a small proportion of the parameters tested showed sex differences at the border of significance. More distinct were seasonal differences, especially in the mitogenic stimulation of lymphocytes and in the number of E and EAC rosetteforming cells; these parameters were highest in June and lowest in November (P 0.1%).

As the blood samples were taken repeatedly from the same animals within two weeks the repeatability of the testing results could be evaluated (Table 1).

TABLE 1   Repeatability of the testing results in pigs.

Parameter tested	Repeatability R
IgM (before immunization)	0.77
IgG (" ")	0.74
IgA (" ")	0.65
Number of leucocytes in the blood	0.50
Percentage of lymphocytes	0.26
" " neutrophiles	0.26
" " eosinophiles	0.14
" " monocytes	0.57
" " basophiles	0.26
" " E-rosetteforming cells	0.56
" " EAC-rosetteforming cells	0.11
" " sIg$^+$ lymphocytes	0.19
Mitogenic stimulation by PHA	0.48
" " " Con A	0.37
" " " PWM	0.72
" " " LPS	0.10
Phagocytosis of yeast cells	0.32
Glucose metabolism during phagocytosis of:	
- killed Staphylococcus epidermidis	0.26
- polystyrene latex particles	0.62

Summing up the results from table1, high repeatability estimates were obtained for the determination of serum immunoglobulin levels, for the percentage of monocytes in the blood, and for the T cell dependent parameters (E rosetteforming cells, mitogenic stimulation by PWM). Low repeatability estimates were found for most parameters of blood cell differentiation and for all B cell dependent parameters (EAC rosetteforming cells, sIg$^+$ cells, mitogenic stimulation by LPS).

Among the significant correlations found the positive correlations between the various mitogenic stimulation assays were most remarkable (for the complete data see Buschmann, 1986). Positive correlations were further observed between the following antibody titres in the immunized animals:

Primary antibody response to tetanustoxoid - primary antibody response
to ovalbumin r = 0.39; secondary antibody response to tetanustoxoid -
secondary antibody response to ovalbumin r = 0.26; primary antibody
response to tetanustoxoid - secondary antibody response to tetanustoxoid
r = 0.55; primary antibody response to ovalbumin - secondary antibody
response to ovalbumin r = 0.69.

The latter results show that there must be some genes which regulate
the antibody-forming capacity non-specifically in pigs. The antibody
response to both antigens was only weakly correlated with the number of
E-rosetteforming cells in the blood (0.15 - 0.18) and was not at all
correlated with the results of the lymphocyte stimulation tests with PHA,
Con A, PWM, and with the serum Ig levels. The B cell dependent stimulation
by LPS, however, was positively correlated with all primary and secondary
antibody titres measured (0.17 - 0.66). As there is no correlation to
the number of B cells in the blood (EAC rosetteforming cells, sIg^+ cells)
the conclusion is justified, that a superior antibody-forming capacity is
not related to the number of B cells but to the reactivity of the single
B cells (LPS stimulation assay).

In the meantime these correlation estimates were repeated after the
testing results of 150 German Landrace x Pietrain pigs had been evaluated
using our new immunocompetence profile including the determination of NK
cell activity and the percentage of lymphocytes reacting with a panel of
monoclonal antibodies obtained from the American Type Culture Collection,
Rockville (Table 2). Again positive correlations were found among the
various lymphocyte stimulation assays. Further a positive correlation
exists between the NK activity and the PHA stimulation rate. Several
positive correlations exist among the results obtained with monoclonal
antibodies (mAb). A positive correlation exists between the NK cell
activity and the labelling of lymphocytes by the mAbs Hb 143 and Hb 147.

The practical value of our immunocompetence scheme was tested in
pigs that had received a whole-body gamma irradiation with a sublethal
dose (250 and 300 rd). In the irradiated animals a retarded beginning
of the antibody response compared with the sham-irradiated controls was
observed. A sublethal irradiation before immunization proved to be more
immunosuppressive than an irradiation with the same dose following
immunization. Further the mitogenic stimulation of lymphocytes was
markedly depressed in the irradiated animals, the B cell dependent

TABLE 2

P E A R S O N   C O R R E L A T I O N   C O E F F I C I E N T S

(n = 150 blood samples)

	Leuco	Lympho	Neutr (mat.)	Neutr (band)	Mono	Yeast-phag.	NBT	Chemil.	PHA	PWM	ConA	LPS	NK	Hb 140	Hb 141	Hb 142	Hb 143	Hb 147
Number of leucocytes	1																	
% Lymphos	0,06	1																
% Neutrophil(mature)	-0,06	-0,87*	1															
% Neutrophil(band)	0,04	-0,29*	-0,08	1														
% Monocytes	-0,08	-0,17	-0,02	0,05	1													
Yeast-phagocytosis	-0,07	-0,11	0,10	0,00	0,01	1												
NB1 - test	0,01	0,08	-0,13	-0,04	0,15	0,09	1											
Chemiluminescence	-0,05	-0,01	0,05	-0,01	-0,11	0,24*	-0,03	1										
PHA	0,28*	-0,05	0,07	-0,01	-0,04	0,02	0,04	0,10	1									
PWM	0,08	-0,08	0,15	-0,14	-0,01	0,23*	-0,12	0,13	0,63*	1								
ConA	0,18	-0,18	0,12	-0,10	-0,10	0,16	0,11	-0,02	0,29*	0,37*	1							
LPS	0,06	0,21	-0,20	0,02	-0,06	-0,09	-0,05	-0,02	0,39*	0,32*	0,10	1						
NK-activity	0,14	0,09	-0,08	0,03	-0,07	-0,10	-0,01	0,22	0,29*	-0,02	0,16	0,11	1					
mAb Hb 140(B+60%T)	-0,01	0,25	-0,35*	0,18	-0,04	-0,05	0,06	-0,14	-0,05	-0,07	-0,23	0,14	-0,09	1				
" Hb 141(T+Ma)	-0,08	0,12	-0,11	-0,05	-0,12	0,12	-0,21	0,04	-0,28	-0,02		-0,01	-0,01	0,00	1			
· " Hb 142(Ma)	-0,06	-0,26*	0,20*	-0,05	0,09	0,08	-0,15	0,02	0,08	0,04	-0,06	-0,29	-0,03	-0,26*	0,51*	1		
" Hb 143(cyt.T)	0,46*	0,11	0,10	-0,01	-0,10	-0,15	0,06	-0,01	0,12	-0,09	0,11	-0,18	0,35*	0,07		-0,21	1	
" Hb 147(T-help.)	0,32*	-0,07	0,35*	-0,13	-0,21	0,06	0,11	-0,13	0,32*	0,13	0,17	-0,32*	0,33*	0,04		-0,04	0,48*	1

stimulation by LPS being extremely radiosensitive. On the other hand the phagocytosis assays showed increased values in the irradiated animals having received 250 rd.

In addition the effect of some adjuvants (incomplete Freund adjuvant, azimexon, glucan, bestatin, BCG) and of the immunosuppressive drug cyclo-phosphamide (endoxan) was tested in pigs by using the immune competence scheme. Clear effects on the antibody-forming capacity and on the mitogenic stimulation rate were observed. The serum immunoglobulin levels of pigs, however, were stable against irradiation and application of immunomodu-lating drugs.

DISCUSSION

Monitoring the immune system becomes increasingly important in veterinary medicine and in animal breeding. For this purpose an immuno-competence profile for swine was developed. It is assumed that the majority of the parameters tested have a direct impact on disease resistance.

Important areas for future research are the detection of gene systems regulating the variation found in the parameters of immunocom-petence. Further the correlation between immunocompetence parameters and production traits must be studied. Additional data on the relative impact of the immunological traits measured on disease resistance are required. The practicability of this approach for improving disease resistance will increase with the number of immunological parameters which can be measured with the precision required.

REFERENCES

Buschmann, H., Kräußlich, H., Herrmann, H., Meyer, J., and Kleinschmidt,A. 1985. Quantitative immunological parameters in pigs - experiences with the evaluation of an immunocompetence profile. Ztschr. Tierz. Züchtungsbiol. 102, 189 - 199.
Buschmann, H. 1986. Which immunological parameters may be used as auxiliary selection criteria for disease resistance in pigs ? Proceedings 3rd World Congress on Genetics Applied to Livestock Production, Lincoln, Nebraska.
Urbaniak, S.J., White, A.G., Barclay, G.R., Wood, S.M., and Kay, A.B. 1978. Tests of immune functions. In: Handbook of Experimental Immunology, 3rd edition (D.M. Weir ed.). Blackwell Scientific Publications.
Winter, M., and Buschmann, H.G. 1987. Measuring phagocytic capacity in polymorphonuclear cells of the pig - a comparison between different assays. J. Vet. Med. B 34,504 - 508.

THE GENETICS OF PARASITE RESISTANCE IN SHEEP

G.A.A. Albers *+ and G.D. Gray*

*Department of Animal Science, University of New England,
Armidale, NSW 2351, Australia
+Euribrid B.V., P.O.Box 30, 5830 AA Boxmeer,
The Netherlands

ABSTRACT

Results are summarized of a series of studies designed to quantify the genetic basis for resistance and resilience in Haemonchus contortus infected sheep and to obtain estimates of productivity responses to selection for these traits. The heritability of resistance to infection was estimated at 0.30–0.40, but the heritability of resilience was too low to allow substantial progress by direct selection for this trait. It was concluded that selection for polygenically controlled resistance would lead to substantial progress for this trait and would also increase productivity of infected animals whilst not affecting productivity in the absence of infection. Moreover, evidence was obtained for the presence of a major resistance gene and it was shown that practical breeding strategies exploiting such a major gene would lead to dramatic improvements of resistance to infection with H. contortus.

INTRODUCTION

Haemonchus contortus is a nematode that parasitizes the abomasum of sheep. By its blood–sucking activity this worm parasite may cause severe anaemia and ultimately death in infected sheep. In those climatic areas of the world in which warm and humid conditions occur simultaneously, H. contortus is the predominant gastro–intestinal parasite of sheep, and production losses caused by this parasite can be very substantial. In some areas of the world profitable sheep farming depends on the use of anthelmintics and in these areas resistance of gastro–intestinal nematodes to the various anthelmintics is developing rapidly (Waller, 1986). Thus there is an urgent need for alternative strategies of helminth control. Since development of a protective vaccine has met with little success so far, exploitation of host genetic resistance could be one approach.

Whether selective breeding for resistance is an attractive alternative or addition to current parasite control methods, ultimately depends on the costs and benefits of such an approach, and the availability of a practicable procedure for measuring resistance in candidate breeding animals.

153

Benefits of a breeding program are expressed in economic units and therefore the effect of selective breeding should be evaluated in terms of reduction of production loss, rather than in increase of resistance to the parasite per se.

Here we distinguish between 'resistance' – the ability to suppress establishment and/or subsequent development of infection, and 'resilience' – the ability to maintain a relatively undepressed production level when infected. The heritability of these characters and their genetic correlations with production traits should determine which one would be the selection criterion to be chosen in the most effective breeding strategy.

The studies presented here were designed to

(1) quantify the genetic basis for resistance and resilience in H. contortus infected sheep, and

(2) quantify the genetic relationships among resistance, resilience and important production characters. Results are evaluated by prediction of direct and correlated responses to selection for these traits and by comparison of different breeding strategies.

MATERIALS AND METHODS

Detailed descriptions of animals used, experimental designs, experimental methods and statistical analyses are given elsewhere (Albers and Gray, 1986; Albers et al., 1987; Albers et al., in press a; Albers et al., in press b; Gray et al., in press).

In short: to estimate heritabilities of, and genetic correlations between, relevant variables, a half–sib design was used. Over 1,000 3–month old Merino lambs, descending from sixty different sires, were tested for resistance and resilience. Resistance was measured by determining faecal parasite egg output and haematocrit after a single dose infection of 11,000 H. contortus larvae. Resilience was measured as the difference between productivity when infected and productivity when uninfected, i.e. depression of production due to infection. Live weight gain as well as wool growth were used as production traits.

RESULTS AND DISCUSSION

Heritability estimates obtained for resistance and resilience traits are given in Table 1. Estimates for resistance characters were moderately high and of the same magnitude as those normally found for production traits, for which selection has been successful in practical sheep breeding programs.

Table 1   Heritability estimates with standard errors for resistance and resilience characters.

Trait		Heritability	St. error
resistance:	faecal egg count	0.30	0.10
	haematocrit	0.40	0.12
resilience:	live weight gain depression	0.09	0.07
	wool growth depression	0.08	0.07

In contrast, the heritability estimates for loss of production due to infection (resilience) were low, indicating that selection for resilience would result in slow rates of progress. Because genetic correlations between resistance and resilience were positive (Albers et al., 1987), selection for resistance would result in progress for resilience as well and vice versa. Due to the much higher heritability for resistance, however, progress for resilience obtained through selection for resistance, would be as high as progress for resilience resulting from direct selection for this trait (Table 2).

Table 2   Relative genetic progress for resistance, resilience and productivity obtained by selection for resistance and resilience (response to direct selection = 100)

Response	Trait	Selection trait	
		Resistance (faecal egg ct)	Resilience (l.w. gain depression)
Resistance	(faecal egg ct)	100	19
Resilience	(l.w. gain depression)	53	100
	(wool growth depression)	182	64
Productivity when uninfected	(l.w. gain)	−27	35
	(wool growth)	−05	−20

A genetically high resistance level would also have the advantage of reducing pasture contamination. Thus, selection for resistance clearly should be preferred to selection for resilience.

Table 2 also shows another important result: selection for resistance (i.e. low faecal egg counts) has a negligible effect on productivity of uninfected animals, so in this respect there are no negative side effects of selective breeding for parasite resistance.

One of the sixty sire groups used in the studies described above displayed an extremely high level of resistance (Albers et al., 1987). Average faecal egg counts for the 17 lambs of this group were 556 eggs per gram at 4 weeks after infection as against 9,033 eggs per gram for offspring of all sixty sires; the resistant sire group clearly was an outlier in the total distribution of sire groups.

In subsequent experiments (Gray et al., in prep) the extremely high level of resistance in offspring from the particular ram was confirmed under conditions of natural infection (expt. A) and when offspring was infected with two different strains of H. contortus larvae (expt. B) (Table 3).

Table 3   Faecal egg counts and haematocrit declines of offspring from the extremely resistant ram and groups of lambs sired by a susceptible sire after naturally acquired infection (expt. A) and after artificial infection with two different strains of H. contortus (expt. B).

Expt.	Sire group	Worm strain	n	Faecal egg ct (epg)	Haematocrit decline (%)
A	resistant	–	49	1,676	2.6
	susceptible	–	49	13,710	7.2
B	resistant	'Kirby'	24	18,784	6.5
	susceptible	'Kirby'	24	38,083	10.2
	resistant	'McMaster'	25	25,237	9.7
	susceptible	'McMaster'	25	40,592	12.4

All differences between resistant and susceptible sire groups, except for the difference in haematocrit decline in McMaster-strain infected lambs, were statistically significant.

It is tempting to think that the extremely resistant sire was the carrier of a major resistance gene. Some evidence for a major gene conferring resistance to H. contortus in sheep was described earlier by Whitlock and Madsen (1958). Also in several mouse models, genetic control of resistance to nematode parasites has been shown or hypothesized to be regulated by only one or a few genes (Wakelin, 1985). However, in our case it appears to be very difficult to confirm the major gene hypothesis through segregation experiments. The lack of sufficient numbers of sheep and the wealth of variability in infection results proves to be a very difficult combination to tackle.

Nonetheless, further research input into this potential major resistance gene seems justified. A sixty per cent increase in resistance of young lambs (expressed as faecal egg counts on a square root scale) would be achieved after one year of using rams homozygous for a dominant resistance gene. In contrast, if resistance would be polygenically regulated with a heritability of 0.29, twelve years of screening would be necessary to achieve only thirty per cent improvement in resistance (Albers and Gray, 1986). However, if simple and cheap procedures to identify genetically superior individuals are available, even such a rate of progress makes selection for resistance to H. contortus an attractive proposition in heavily infected areas.

ACKNOWLEDGEMENT

The work presented in this paper was supported by the Australian Meat Research Committee and the Wool Research Trust Fund.

REFERENCES

Albers, G.A.A. and Gray, G.D. 1986. Breeding for worm resistance: a perspective. In "Parasitology Quo Vadit?" (Ed. M.J. Howell). (Austrialian Academy of Science, Canberra). pp. 559-566.

Albers, G.A.A., Gray, G.D., Piper, L.R., Barker, J.S.F., Le Jambre, L.F. and Barger, I.A. 1987. The genetics of resistance and resilience to Haemonchus contortus infection in young Merino sheep. Int. J. Parasit, 17, 1355-1363.

Albers, G.A.A., Gray, G.D., Le Jambre, L.F., Piper, L.R., Barger, I.A. and Barker, J.S.F. In press a. The effect of Haemonchus contortus infection on live weight gain and wool growth in young Merino sheep. Austr. J. Agric. Res.

Albers, G.A.A., Gray, G.D., Le Jambre, L.F., Piper, L.R., Barger, I.A. and Barker, J.S.F. In press b. The effect of Haemonchus contortus infection on haematological parameters in young Merino sheep and its significance for productivity. Anim. Prod.

Gray, G.D., Albers, G.A.A., Burgess, S.K., Le Jambre, L.F. and Windon, R.G. In prep. Response of genetically resistant Merino lambs to artificial and naturally acquired infections of Haemonchus contortus.

Wakelin, D. 1985. Genetic control of immunity to helminth infections. Parasit. Today 1, 17–23.

Waller, P.J. 1986. Anthelmintic resistance in sheep nematodes. Agric. Zool. Rev. 1, 333–373.

Whitlock, J.H. and Madsen, H. 1958. The inheritance of resistance to trichostrongylidosis in sheep II. Observations on the genetic mechanism in trichostrongylidosis. Corn. Vet. 48, 134–145.

THE BIOZZI MODEL APPLIED TO THE CHICKEN

A.J. van der Zijpp and M.G.B. Nieuwland

Department of Animal Husbandry, Agricultural University,
P.O. Box 338, 6700 AH Wageningen, The Netherlands

SUMMARY

Resistance to infectious diseases is influenced by environmental factors as well as genetic constitution. Qualitative and quantitative inheritance of resistance is known for a variety of poultry diseases. However, genes directly responsible for resistance or susceptibility have not yet been identified. In this paper the direct and correlated responses to selection for antibody production to sheep red blood cells (SRBC) are presented and the strategy to identify the genes responsible for antibody production to SRBCs will be discussed.

After seven generations of selection for antibody production to SRBCs we have obtained high (H) and low (L) antibody titer producing lines of ISA Warren origin. Differences in antibody production (titer 7.1 versus 2.1 for H and L lines in the seventh generation) became significant (P<0.5) from the first selection onwards. Upon contact exposure to Marek's disease virus the H line showed 20-30% less mortality than the L line (P<0.001).

The divergence in H and L lines demonstrates that genetic variation in immune responsiveness and disease resistance exists and exemplifies genetic differences between breeds and crosses in poultry production. As has previously been demonstrated in the mouse, genes contributing to these immunological and pathological differences between H and L lines may belong to polymorphic immunoglobulin (Ig) loci and/or loci of the major histocompatibility complex (MHC). To establish the role of these loci in the chicken, F_1 and F_2 generations will be obtained by crossing the H and L lines. In the F_1 and F_2 hybrids background genes of H and L lines will be randomly distributed. With the application of serological techniques for identifying Ig-allotypes and MHC-alleles it will then become possible to determine the effects of the individual alleles on SRBC antibody production and resistance to Marek's disease.
Keywords: Selection lines, B-complex, Marek's disease, SRBC.

INTRODUCTION

Principal methods to control infectious diseases include hygienic measures, effective management of housing, climate and nutrition, preventive medication, vaccination, eradication and breeding for genetic resistance. Breeding for resistance may be preferred, because it diminishes the need for continuous or repeated preventive measures and it is conferred to following generations.

A detailed treatise on genetic variation in immune responsiveness and disease resistance was presented by Hartmann (1982) and Van der Zijpp (1983). Recent studies emphasize the interactions of the immune system in the defense against disease (Powell, 1987). Resistance is the result of a complex reaction composed of innate and acquired immunity. Molecules of the chicken major histocompatibility complex (B-complex) play a central role in the immune defense. In a detailed review Bacon (1987) has indicated that chickens can be typed for disease resistance traits on the basis of their polymorphic B-haplotypes.

In many situations genetic resistance in combination with several preventive methods leads to the best results. Gavora & Spencer (1983) studied the effect of combined vaccination and selection on Marek's disease resistance. Maximum resistance to Marek's disease, expressed in percentage mortality, was found in vaccinated chickens of the most resistant strain; for maximum productivity, however, vaccination had to be applied in a strain selected for production. This implies that breeders have to consider the pathogenetic environment and the opportunities for effective vaccination to define their breeding goals with respect to diseases. For some diseases, such as Marek's disease and coccidosis, the present interest in genetic resistance is based on availability of vaccination or medication procedures and their effectiveness in diverse geographical locations and a variety of stocks.

Breeding for disease resistance is based on pathological and immunological traits and on genetically linked markers. These traits and markers have to show genetic variation, have to be relevant for disease resistance or susceptibility and have to be determined easily in routine testing on large numbers of animals. In mice extensive selection studies have been carried out by Biozzi et al. (1979). These authors concluded that antibody production was negatively correlated with microbicidal activity of macrophages, but cell-mediated immunity was not affected. This pattern determines their effective resistance to a variety of diseases.

In this study we present results obtained from the selection of chicken lines for high (H) and low (L) antibody production to SRBCs, demonstrating a significant effect on resistance to Marek's disease. Using immunological and biotechnological techniques in association studies, the effects of particular loci and genes on antibody production and Marek's disease mortality can now be established.

CHARACTERIZATION OF HIGH AND LOW ANTIBODY PRODUCING LINES

Selection for antibody production to complex antigens to which animals have not previously been exposed, may provide a tool to breed for disease resistance. In mice, lines have been selected for antibody production to complex antigens and tested extensively for immunological and pathological differences (Biozzi et al., 1979). Selection for high antibody production resulted in diminished microbicidal activity of macrophages, while cell-mediated immunity remained unaffected. Pathological differences became obvious through effective resistance of lines to a variety of diseases.

In 1980 we started a selection experiment using hybrid ISA Warren chickens. Chicks were intramuscularly injected at 37 days of age with SRBCs, and total hemagglutinating antibody titers were determined five days post immunization. Results of selection for high and low antibody production, as well as for a randombred control (C) line after seven generations are shown in Table 1 and Figure 1.

<u>Table 1</u>. Titers of SRBC-agglutinating antibodies in generation seven of control (C) and selected high (H) and low (L) line chickens.

Sex	Line		
	C	H	L
Male	4.59 (n = 176)	6.79 (n = 123)	1.88 (n = 188)
Female	5.23 (n = 172)	7.37 (n = 125)	2.39 (n = 189)

In the base population, a heritability of hemagglutinating antibody titers for half sibs was estimated at .57 ± .21. Realised heritabilities (h^2) were calculated after seven generations of selection by regression of cumulative response on cumulative selection differential (Table 2). In the L line, realised heritabilities are closest to the half sib estimate in the base population.

The frequency distribution of chicks is shown in Figure 2. In each generation the overlap between the H, L and C line decreases, signifying the response to selection.

The immunological characterization of the differences between the lines (Van der Zijpp & Nieuwland, 1986) has produced significant differences in the humoral response against diverse antigens between lines

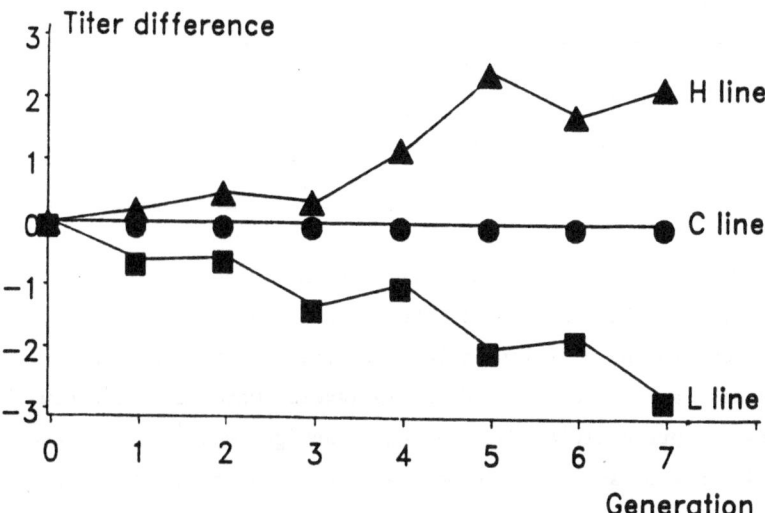

SRBC antibody titer (Selection result)

Figure 1. Selection response in H and L lines for SRBC antibody titers relative to a random control (C) line.

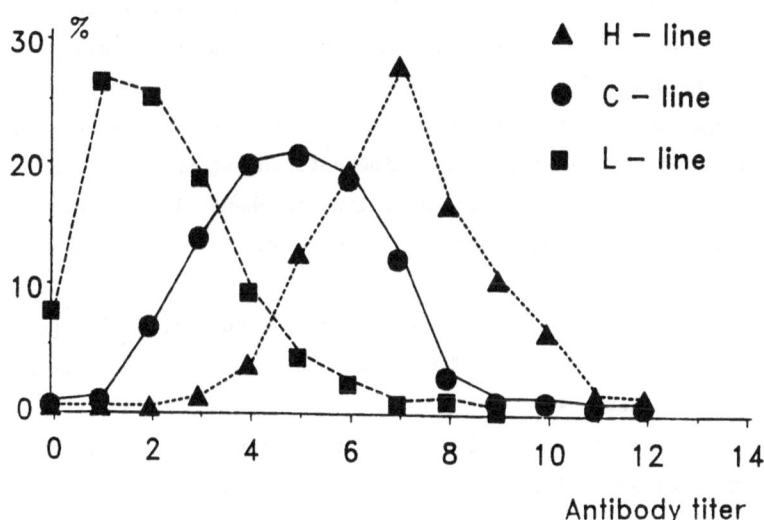

Day 5 Titer Frequencies per Line — S7

Figure 2. Frequency distribution of H, L and C lines after seven generations of selection.

Table 2. Realised heritabilities for haemagglutinating SRBC antibody titers
based on seven generations of selection of high (H) and low (L)
lines.

Line	h^2	Males	Females
H	.16	.22	.13
L	.22	.31	.17

(H > L). This difference is also reflected in the number of haemolytic
plague forming cells, in favor of the H line. Tests for cell-mediated
immunity revealed no differences between lines. Differences in
phagocytosis, antigen uptake and enzyme levels, were small and usually not
significant. Intracellular degradation of antigen and surface presentation
and persistence of antigen have not yet been studied, but may contribute
significantly to the overall difference between lines.

Differences in productivity between the lines have been observed in
the sixth generation. In Table 3 production data have been summarized.
Except for egg weight the H and L line have been less productive compared
to the C line. The L line, in comparison with H, has gained in bodyweight,
but reduced the total number of eggs very significantly. Data of previous
generations have not yet been analysed, so it is not known whether these
data confirm trends over generations. A distinct negative genetic
correlation was found in the base-population : r_g = -.57 for bodyweight at
57 days of age and primary titer where as r_p was -.06. In all following
generations the body weight of the H line chickens at that age was
significantly reduced compared to the L line chickens. In contrast, the H
line chickens have a higher relative spleen weight before and after
immunization.

Disease resistance has been determined in a contact challenge
experiment for Marek's disease. The percentage mortality for H and L line
birds is shown in Table 4.
Differences between sexes and lines were significant (P<0.001). Preliminary
results have indicated significant differences in the frequency of
B-complex alleles between the lines.
Although data of the characteristics of H and L lines are accumulating
slowly, they apparently show many similarities with the selection lines of

Siegel and Gross at Virginia Polytechnic Institute and State University
(Siegel & Gross, 1980; Siegel et al., 1982; Dunnington et al., 1986).
Despite different breeds (White Leghorn versus ISA Warren), effects on
immune responsiveness, Marek's disease resistance and body weight were
similar.

Table 3. Production data of lines selected for antibody production for 6
generations.

	Line		
Trait	H	C	L
Bodyweight hens (g) at 126 days of age	1749[b]	1797[a]	1792[a]
Age at first egg (days)	142.3[ab]	141.3[a]	143.6[b]
Total number of eggs	211.6[a]	213.4[a]	198.6[b]
Egg weight (g)	51.7[a]	48.8[b]	49.3[b]

a, b : significantly different P < .05

Table 4. Percentage mortality measured 15 weeks after exposure of Marek's
disease virus to high (H) and low (L) lines.

Line	number		Marek's disease mortality	
	males	females	males	females
H	100	92	13	28
L	125	130	33	59

LOCI AND GENES INVOLVED IN ANTIBODY PRODUCTION AND MAREK'S DISEASE
RESISTANCE

Biozzi et al. (1979) estimated that about ten independent loci
controlled the antibody response to SRBCs in mice. The inter-line
difference could be explained only in part by two defined loci. Positive
associations were demonstrated with an allotypic marker of the Ig heavy
chain and with certain alleles of the mouse MHC (H-2). The Ig locus
explained 10% and the H-2 complex explained 20% of the difference in
agglutinin synthesis.

Association studies were extended after the discovery of large, or even absolute differences in frequency of Ig allotypes and H-2 alleles between the H and L lines. F_1 and F_2 hybrids, as well as backcrosses between F_1 hybrids and H or L parental lines were produced. The normal immunization procedure was carried out and hemagglutination titers to SRBCs determined. Maximum differences in agglutinin titer between Ig allotypes and H-2-haplotypes were 10 and 20% respectively. The inter-line divergence may be determined directly by these loci or by closely linked loci.

In chickens extensive polymorphism has been described for IgG C_H and IgM C_H allotypes (Benedict & Berestecky, 1987). Fourteen allotypic specificities have been described for IgG C_H and five for IgM C_H, which belong to the G-1 and M-1 loci respectively.
The B-complex encompasses three tightly linked loci which code for B-F (class I), B-L (class II) and B-G (class IV) molecules. The B-G molecules are highly polymorphic products of genes closely linked to the class I and II genes. Polymorphism of B-G molecules can easily be defined using typing sera in a direct hemagglutination test. For example, the thus defined B-G21 allele is associated with resistance to Marek's disease. However, resistance is rather influenced by the B-F21 or B-L21 locus than by the B-G21 locus (Briles et al., 1983; Crone & Simonsen, 1987).

Based on preliminary results, we expect to find distinct differences of the B-haplotype distribution between the H and L chicken lines. However, a complicating factor, when compared to the mouse experiments, is that four to six haplotypes may be present in both lines. We will therefore proceed to evaluate the effects of these haplotypes in H, L and C lines on antibody production and morbidity caused by Marek's disease with a mixed model now (Bentsen and Larsen, 1988). The results obtained in the F2 generation, an evaluation of B-G haplotypes against random background genes, can then be compared with the statistically obtained estimates.

CONCLUSIONS

Selection for antibody production to SRBC in the chicken has been succesfull. Correlated responses sofar appear to be limited to effects on humoral response to other antigens and to spleen size. A challenge to Marek's disease virus resulted in highly significant differences in favor of the H line. Research will now be aimed at estimation of effects of particular loci, on differences in antibody production and Marek mortality. If these gene complexes play an important role molecular identification of these genes for transfection studies becomes feasible.

ACKNOWLEDGEMENT

The immunological research and selection of lines was performed by Mr. M.G.B. Nieuwland. Ir. M.B. Kreukniet was responsible for the statistical analysis of the immune response and production data.

REFERENCES

Bacon, L.D., 1987. Influence of the major histocompatibility complex on disease resistance and productivity. Poultry Science 66: 802-811.

Benedict, A.A. & J.M. Berestecky, 1987. Special features of avian immunoglobulins. In: A. Tovianen & P. Tovianen (Ed.): Avian immunology: basis and practice. Volume I, CRC Press, Inc., Boca Raton, Florida. p. 113-126.

Bentsen, H.B. and H.J. Larsen, 1988. MHC-associated effects in laying hens on immunological parameters in blood serum. Abstracts 21st International Conference on Animal Blood Groups and Biochemical Polymorphisms, Torino, Italy. 4-8 July 1988.

Biozzi, G., D. Mouton, O.A. Sant'Anna, H.D. Passos, M. Gennari, M.H. Reis, V.C.A. Ferreira, A.M. Heumann, Y. Bouthillier, O.M. Ibañez, C. Stiffel & M. Sigueira, 1979. Genetics of immunoresponsiveness to natural antigens in the mouse. Current Topics in Microbiology and Immunology 85: 31-99.

Briles, W.E., R.W. Briles, R.E. Taffs & H.A. Stone, 1983. Resistance to a malignant lymphoma in chickens is mapped to a subregion of major histocompatibility (B) complex. Science 219: 977-979.

Crone, M. & M. Simonsen, 1987. Avian Major Histocompatibility Complex. In: A. Toivanen & P. Toivanen (Ed.): Avian Immunology: basis and practice. Volume II. CRC Press, Inc., Boca Raton, Florida. p. 25-42.

Dunnington, E.A., A. Martin, W.E. Briles, R.W. Briles & P.B. Siegel, 1986. Resistance to Marek's disease in chickens selected for high and low antibody responses to sheep red blood cells. Archiv für Geflügelkunde.

Gavora, J.S. & J.L. Spencer, 1979. Studies on genetic resistance of chickens to Marek's disease-a review. Comp. Immun. Microbiol. infect. Dis. 2: 359-371.

Gavora, J.S. & J.L. Spencer, 1983. Breeding for immune responsiveness and disease resistance. Animal Blood Groups and Biochemical Genetics 14: 159-180.

Hartmann, W., 1982. Selection and genetic factors on resistance to disease on chickens. Proceedings 2nd. World Congress on Genetics applied to Livestock Production, Madrid, Spain. p. 709-726.

Powell, P.C., 1987. Immune mechanisms in infections in poultry. Veterinary Immunology and Immunopathology 15: 87-113.

Siegel, P.B., W.B. Gross & J.A. Cherry, 1982. Correlated responses of chickens to selection for production of antibodies to sheep erythrocytes. Animal Blood Groups and Biochemical Genetics 13: 291-297.

Siegel, P.B. & W.B. Gross, 1980. Production and persistence of antibodies in chickens to sheep erythrocytes. I. Directional selection. Poultry Science 59: 1-5.

Vaiman, M., P. Chardon & D. Cohen, 1986. DNA polymorphism in the major histocompatibility complex of man and various farm animals. Animal Genetic 17: 113-133.

Zijpp, A.J. van der, 1983. Breeding for immune responsiveness and disease resistance. World's Poultry Science Association Journal 39: 118-131.

Zijpp, A.J. van der, 1984. Heritability for the immune response to Newcastle disease virus vaccine and relationship with production traits. Proceedings of the 17th World's Poultry Congress, Helsinki, Finland. p. 518-520.

Zijpp, A.J. van der & M.G.B. Nieuwland, 1986. Immunological characterisation of lines selected for high and low antibody production. Proceedings 7th European Poultry Conference, Paris, France. p. 211-215.

GENETIC RESISTANCE TO BOVINE MASTITIS

P. Madsen

National Institute of Animal Science
Research Center Foulum
P.O. Box 39, 8830 Tjele
Denmark

ABSTRACT

The genetic background for the incidence of and resistance/susceptibility to mastitis is briefly reviewed.

A Danish research project on genetic resistance to mastitis is described. The main results show that the heritability for incidence of clinical mastitis is low (0.03), that clinical mastitis and somatic cell count are highly genetic correlated (0.81) and that the genetic correlation between incidence of clinical mastitis and butter fat yield is unfavourable (0.66). Several teat traits have favourable genetic correlations with incidence of clinical mastitis.

Selection strategies including clinical mastitis, somatic cell count and teat traits in different combinations with butter fat yield showed a lesser increase in clinical mastitis than strategy based only on butter fat yield. However, this was at the expense of genetic progress in yield. Constraining the strategies to keep mastitis at its present level further reduced the genetic progress in yield. Genetic reduction of the incidence of clinical mastitis is possible, but at a large expense in yield.

Marker genes of the β-lactoglobulin system and the M blood groups system showed an effect on incidence of clinical mastitis. Elimination of the susceptible genes is not recommended because of lack of knowledge about effects on other traits.

INTRODUCTION

Mastitis is a widespread multifactorial bovine disease of great economical importance. In Denmark 30-40% of all veterinary treatments of dairy cows are due to mastitis. In most countries with a developed dairy industry mastitis is one of the major reasons for culling of cows.

Traditionally, mastitis controlling programs have focused on managemental and environmental factors. Such programs, together with adequate therapy, have been effective in reducing the incidence of some types of mastitis. This was seen as a shift in major causative organisms, that has occurred during recent decades (e.g. Schalm et al., 1971; Dood, 1985). However, the effect on the mastitis problem as a whole has been limited.

Several studies have shown that genetic variation in mastitis resistance/susceptibility exists. For a review of the literature until 1979 see Lie et al. (1980). More recent Solbu (1984), Emanuelson (1987) and Madsen et al. (1987) found significant genetic variation in the incidence of clinical mastitis, but heritabilities were very small, from 1% to 5%.

Unfavourable genetic correlations between incidence of clinical mastitis and milk production traits have been reported by Wilton et al. (1972), Bunch et al. (1984), Emanuelson (1987) and Madsen et al. (1987).

DEFENCE MECHANISMS AGAINST MASTITIS

The cow has general defence mechanisms against infectious diseases and the mammary gland has specific defence mechanisms related to mastitis resistance/susceptibility.

The general defence mechanisms are related to the actual status of the cow, and it is known that mastitis often occurs in connection with stress or other diseases e.g. milk fever, ketosis and other digestion disorders.

The local defence mechanisms of the mammary gland can be divided into passive and active mechanisms as shown in fig. 1.

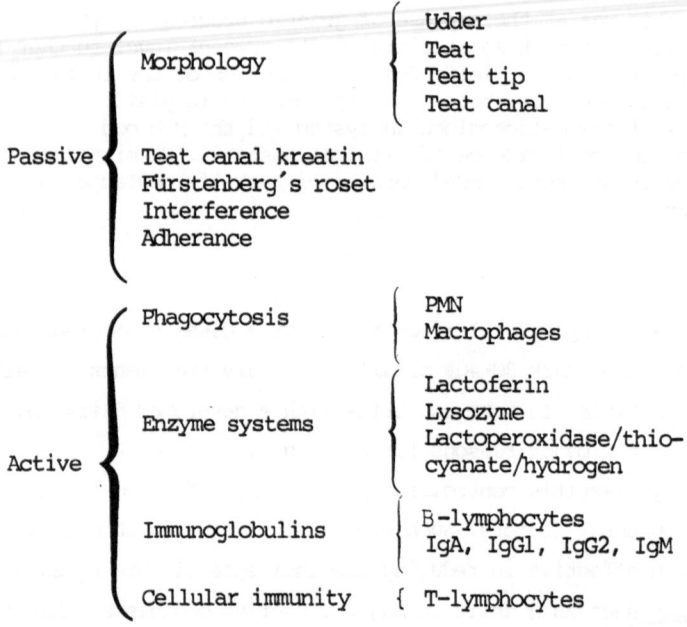

Fig. 1   Local defence mechanisms of the mammary gland against mastitis.

It is known from the literature that several of the passive defence mechanisms show genetic variation. Especially the morphological traits have relative high heritabilities (see review by Lie et al. (1980) and by Seykora and McDaniel (1985)).

More of the active defence mechanisms, as for instance the content of phagocytising cells, enzyme systems and the concentration of antibodies are also known to be genetically influenced (for a review see Lie et al. (1980) and Miller (1982)).

MARKERS FOR MASTITIS RESISTANCE/SUSCEPTIBILITY

The different defence mechanisms can be regarded as quantitative indicator traits for mastitis resistance/susceptibility. But the active defence mechanisms may be regarded as indicators of both resistance and disease. Evidently, this circumstance affects the potential applicability of such parameters in breeding for mastitis resistance.

Traits with simple inheritance such as blood group systems, polymorphic protein systems in blood and milk as potential marker genes for mastitis resistance/susceptibility have been studied by Mitscherlich et al. (1966) and Larsen et al. (1985). Lately studies have included the major histocompalibility complex (BoLA). Solbu et al. (1982) found an association between BoLA wl6 and susceptibility, and BoLA w2 and restistance to mastitis. Meyer et al. (1984) found association between mastitis susceptibility and BoLA w6 together with wl3.

DANISH RESEARCH ON GENETIC RESISTANCE TO MASTITIS

In 1978 a comprehensive research project was initiated. The aim of the project was to produce the necessary genetic knowledge to decide whether or not mastitis resistance ought to be included in the breeding programs.

The investigation was carried out in 67 herds of Red Danish dairy cattle during the period from August 1978 to December 1982. A total of 1344 cows sired by 124 sires were fully involved in the investigation as "experimental cows" while the remaining cows in the herds participated partly only. The experimental cows initiated 1344, 1045, 628 and 385 1st, 2nd, 3rd and 4th plus later lactation respectively. Experimental cows plus the remaining cows (in the following named "all cows") initiated 4767, 3720, 2290 and 2398 1st, 2nd, 3rd and 4th plus later lactation respectively. The plan for yield, disease, conformation recordings and blood and milk sampling is outlined in table 1.

TABLE 1   Experimental recordings, milk and blood sampling, objective and frequency.

Sample/recording	Objective	Frequency
Quarter milk samples (Experimental cows)	Mastitis diagnostics, based on bacteriological examination, somatic cell count and concentration of BSA	Every 3rd month
Disease recordings (All cows)	Recording of clinical mastitis and other clinical diseases	Daily
Milk samples (All cows)	SCC as a potential indicator for mastitis resistance	Monthly
Mild yield (All cows)	Correlation between yield mastitis, SCC etc.	Monthly
Blood samples (Experimental cows)	Blood and serum protein types, conc. of immunoglobulins, albumin and total protein in serum as potential marker genes	Once/cow
Milk samples (Experimental cows)	Milk protein types as potential marker genes	Once/cow
Udder and teat traits (Experimental cows)	Correlation between udder/teat morphology and mastitis	Once per lactation
Milking performance (Experimental cows)	Correlation between milking performance and mastitis	Once per lactation

The project has been reported in details by Madsen et al. (1987). The main results based on 1st lactation data concerning clinical mastitis, milk production capacity, and some indicator traits and marker genes as well as the effect of different selection strategies are presented in the following.

Traits

The incidence of clinical mastitis (CLM), defined as the number of veterinary treated cases of mastitis within lactation. Treatments of the same quarter within fourteen days were not considered as a new case.

Milk production capacity was the 305-days lactation butter fat yield (BF305). The daily excretion of somatic cells, log_{10}-transformed (LDSC), distance floor/teat tip, front (DFTT), distance between front teats

(DRFLF), length of front teats (TL) and total milking time (TMT) were used
as indicator traits for mastitis resistance/susceptibility.

Marker genes

As marker genes 11 blood group systems (A, B, C, F, J, L, M, S, Z, R'
and T'), 5 blood protein-enzyme systems (Transferrin, Amylase, Ceruloplas-
min, Carbonanhydrase and Adenosindeaminase), and 4 milk protein systems
(β-lactoglobulin, α_{s1}-, β-, and κ-casein) were used.

Statistical methods

The phenotypic and genetic parameters for CLM, BF305 and the indicator
traits were estimated by an iterative multi trait resticted maximum likelihood (REML) procedure. The model used included the effects of herd, year-
month of calving, age at calving and sire of the cow.

The effects of the marker genes on CLM were estimated simultaneously
by least squares as proposed by Mather (1977). The model used included the
effects of herd, year-month of calving, age at calving and the effects of
all the marker systems.

Results

Phenotypic and genetic parameters for CLM, BF305 and the indicator
traits are given in table 2.

The results show that the heritability of CLM is low (0.03), while the
indicator traits have heritabilities in the range from 0.13 (LDSC) to 0.44
(DRFLF). A high positive genetic correlation (0.81) between CLM and LDSC
was found. Thus, high cell count is genetically linked to higher incidence
of clinical mastitis. A strong positive genetic correlation (0.66) was
found between BF305 and CLM, which is unfortunate since selection for yield
would result in a genetic determined increase in the incidence of clinical
mastitis. The genetic correlations between CLM and the teat traits were advantageous and numerically in the range from 0.13 to 0.24.

Results from the analysis showed that only few of the included potential marker genes exhibited any effect on CLM. Cows carrying the genotype
AA of β-lactoglobulin were more frequently affected by clinical mastitis
than those of genotype BB. In 1st lactation the difference in CLM between
genotype AA and BB was 0.32. Cows lacking the M' blood group factor showed
a lower incidence of clinical mastitis than cows with the M' blood group
factor. The difference in CLM in 1st lactation was 0.19.

TABLE 2   Means, phenotypic (s_P) and genetic (s_A) standard deviations, heritabilities (h^2) and phenotypic (r_P) and genetic (r_A) correlations for incidence of clinical mastitis (CLM), 305-days butter fat yield (BF305), average daily cell excretion, $\log_{10}$-transformed (LDSC), distance floor/teat tip, front (DFTT), distance between front teats (DRFLF), teat length, front (TL) and total milking time (TMT).

Trait	Mean	s_P	s_A	h^2, r_P and r_A[1) CLM	BF305	LDSC	DFTT	DRFLF	TL	TMT
CLM	0.47	1.06	0.19	0.03	-0.01	0.25	-0.07	0.04	0.10	0.06
BF305	210.2	35.1	14.6	0.66	0.17	0.10	-0.27	0.18	-0.03	0.13
LDSC	5.197	0.39	0.14	0.81	0.60	0.13	-0.20	0.09	0.11	0.03
DFTT	47.8	3.19	1.80	-0.24	-0.07	-0.40	0.32	-0.12	-0.28	-0.14
DRFLF	18.6	3.26	2.16	0.13	-0.02	0.21	-0.20	0.44	0.06	0.14
TL	5.8	0.94	0.51	0.24	0.10	0.30	-0.59	0.22	0.30	0.13
TMT	5.62	1.64	0.73	-0.06	-0.10	-0.05	-0.04	0.38	0.11	0.20

1) h^2 on the diagonal, r_P above and r_A below. All mutual correlations between the teat traits and milking time are from Jensen (1985).

Selection strategies

Selection for improved mastitis resistance may either be direct selection towards a lower incidence of clinical mastitis or by indirect selection based on indicator traits or marker genes.

The expected effects of different selection strategies among bulls on the incidence of clinical mastitis, butter fat yield, teat traits and milking time are given in table 3.

The results in table 3 show that one-sided selection for increased butter fat yield (strategy 1) would cause a genetic increase in the incidence of clinical mastitis of 40%. Using strategy 2, which is an approximation of the present breeding strategy in Denmark would genetically increase the level of clinical mastitis by 33%. Extending this strategy by including clinical mastitis and/or somatic cell count (strategies 3 to 6) would result in a rather limited effect on the genetically conditioned increase in the incidence of clinical mastitis.

The results from strategies 2 to 6 with constraints on the incidence of clinical mastitis show that it is possible to keep the genetic level for clinical mastitis at its present level and still improve mild yield capacity, or though at a slower rate. The loss in genetic progress in yield may be considerably reduced, if clinical mastitis and/or somatic cell count is included in the selection index.

TABLE 3   Effect of different selection strategies after selection
with a selection differential corresponding to 10% progress when selection for butter fat only. The traits included are: incidence of
clinical mastitis (CLM), 305-days butter fat yield (BF305), average
daily excretion of somatic cells, $\log_{10}$-transformed (LDSC), distance
floor/teat tip, front (DFTT), distance between front teats (DRFLF),
length of front teats (TL) and total milking time (TMT).

Stra-	Traits in aggregate genotype and/or in index[1]							Relative level		
tegy	BF305	DFTT	DRFLF	TL	TMT	LDSC	CLM	CLM	BF305	BF305[2]
1	A,I							140.3	110.0	-
2	A,I	A,I	A,I	A,I	A,I			132.5	109.3	102.9
3	A,I	A,I	A,I	A,I	A,I		A	127.5	108.5	102.9
4	A,I	A,I	A,I	A,I	A,I	I	A	125.6	108.4	104.6
5	A,I	A,I	A,I	A,I	A,I		A,I	125.6	108.4	104.6
6	A,I	A,I	A,I	A,I	A,I	I	A,I	124.9	108.4	105.0
7							A,I	54.9	94,9	-
EW[3]	58/kg	70/cm	-96/cm	-157/cm	-88/min		-1500/case			

1) A indicate that the trait is included in the aggregated genotype,
   I indicate that the trait is used as source of information in the index.

2) Constant incidence of clinical mastitis constrains these indeces.

3) Economic weight, Dkr./unit.

Strategy 7 (one-sided selection for lower incidence of clinical mastitis) shows further that the incidence of clinical mastitis can be reduced
considerably by selection, but dairy merit would then decrease.

Among marker genes the A-gene in the β-lactoglobulin system and the
M′-gene in the M blood group system was most promising. The effect of eli-
minating these genes would be a decrease in clinical mastitis in 1st lactation by 8.8% and 12.3% respectively. Knowledge of the effect of these genes
on other traits of importance, as well as the linkage of the M blood group
system to the BoLA system (Leveziel and Guèrin, 1980; Dam et al., 1984; Leveziel and Hines, 1984), is still inadequate. Hence, using these genes as
markers for mastitis resistance/susceptibility is not recommendable presently.

REFERENCES

Bunch, K.J., D.J.S. Heneghan, K.G. Hibbit & G.J. Rowlands, 1984. Genetic influences on clinical mastitis and its relationship with milk yield, season, and stage of lactation. Livest. Prod. Sci. 11:91-102.

Dam, L., B. Larsen & H. Østergård, 1985. Relations between BoLA and M blood groups in cattle. Abstract of paper presented at the 19th International Conference on Animal Blood groups and Biochemical Polymorphisms (Göttingen, 1984). Animal Blood Groups and Biochemical Genetics 16, suppl., 1:75.

Dodd, F.H., 1985. Progress in mastitis control. Kieler Milchwirtschaftliche Forschungsberichte 37, 216-223.

Emanuelson, U., 1987. Genetic studies on the epidemiology of mastitis in dairy cattle. Rep. No. 73, Swedish University of Agricultural Sciences, Dep. of Anim. Breeding and Genetic. Thesis.

Jensen, J., 1985. Avlsværdivurdering af tyre for eksteriøregenskaber. Rep. No. 595, Nat. Inst. Anim. Sci., Copenhagen, 116 pp.

Larsen, B., N.E. Jensen, P. Madsen, S.M. Nielsen, O. Klastrup & P. Schmidt Madsen, 1985. Association of the M blood group system with bovine mastitis. Anim. Blood Groups and Biochemical Genetics, 16:165-173.

Leveziel, H. & G. Guèrin, 1980. An attempt to detect genetic linkage between the major bovine histocompatibility system and other genetic markers. Abstract of paper presented at the 17th International Conference on Animal Blood groups and Biochemical Polymorphisms (Wageningen, 1984). Animal Blood Groups and Biochemical Genetics 11, suppl., 1:28.

Leveziel, H. & H.C. Hines, 1984. Linkage in cattle between the major histocompatibility complex (BoLA) and the M blood group system. Gènètique, Sèlection, Evolution, 16:405-416.

Lie, Ø., P. Madsen & E. Persson, 1980. Mastitt hos storfe. Resistensmekanismer, spesielt fra et avlsmessig synspunkt. Rapport, Nordisk Kontaktorgan for Jordbrugsforskning (NKJ), 42 pp.

Madsen, P., S.M. Nielsen, M.D. Rasmussen, O. Klastrup, N.E. Jensen, P.T. Jensen, P.S. Madsen, B. Larsen & J. Hyldgaard-Jensen, 1987. Undersøgelser over genetisk betinget resistens mod mastitis. Rep. No. 621, Nat. Inst. Anim. Sci., Copenhagen, 227 pp.

Mather, R.E., 1977. Review of regional project NE 62: Relationship between genetic markers and performance in dairy cattle. J. Dairy Sci., 60:482-492.

Meyer, F., S. Cwik, G. Erhardt, D.O. Schmid & B. Senft, 1984. Zum Nachweis von Lymphozytenantigenen des BoLA-Systems bei Rindern mit Sekretionsstorungen. Zuchtungskunde 56, 108-114.

Miller, R.H., 1982. Genetics of resistance to mastitis. Proceedings of the Second World Congress on Genetics Applied to Livestock Production 5, 186-198.

Mitscherlich, E., F. Hogreve, J. Koch & E. Scupin, 1966. Untersuchungen über Beziehungen zwischen Mastitisresistenz und Blutgruppenfaktoren beim schwarzbunten Niederungsrind. Deutsche Tierärztliche Wochenschrift, 73:97-102.

Schalm, O.W., E.J. Carroll & N.C. Jain, 1971. Bovine mastitis. Lea and Febiger, Philadelphia.

Seykora, A.J. & B.T. McDaniel, 1985. Udder and teat morphology related to mastitis resistance: A review. J. Dairy Sci., 68:2087-2093.

Solbu, H., 1984. Disease recording in Norwegian dairy cattle. II. Heritability estimates and progeny testing for mastitis, ketosis and "all diseases". Z. Tierzüchtg. Züchtgsbiol. 101:51-58.

Solbu, H., R.L. Spooner & Ø. Lie, 1982. A possible influence of the bovine
    major histocompatibility complex (BoLA) on mastitis. Proceeding of the
    Second World Congress on Genetics Applied to Livestock Production,
    Madrid, 7:368-371.
Wilton, J.W., L.D. Van Vleck, R.W. Everett, R.S. Guthrie & S.J. Roberts,
    1972. Genetic and environmental aspects of udder infections. J. Dairy
    Sci. 55:183-193.

SESSION 6

General discussion

Chairpersons: Dr. W. Sybesma and Dr. A.J. van der Zijpp

GENERAL DISCUSSION AND CONCLUSIONS

The seminar "Reducing the costs of disease by breeding for disease resistance" provided an overview of mainly European research on genetic resistance to disease. In the first three sessions the Major Histocompatibility Complex, with major emphasis on genetic structure and techniques for typing, was discussed for fish, poultry, swine, cattle and horses. The fourth session was devoted to associations between the MHC and diseases. In the fifth session a series of papers was presented on selection for immune response parameters in pigs and poultry, resistance versus resilience for parasitic infections in sheep and genetic aspects of mastitis. During the closing session the following items were discussed with respect to future research policies and strategies.

Species. In comparison to man and mouse the study of the MHC of farm animals is still in its infancy. The knowledge of the MHC is most advanced in cattle and poultry, both with regard to genetic structure and function e.g. T-cell-restriction of the MHC. Fish are the least advanced in this sense but have high potential for experimental and practical applications. Work in progress in Danish and Dutch laboratories shows that an MHC probably also exists in fish, because allograft responsiveness has been correlated with serologically detectable cell surface antigens. A biochemical characterization of the putative MHC-molecules hopefully will present the final proof. In other species efforts are continuing to confirm the presence of class II products, and to characterise the genetic structure of class I and II regions.

Methods. In most species, research in MHC polymorphism involves serological identification with allo-antisera, gene product characterization with SDS-PAGE, 1- and 2-dimensional iso-electric focussing (IEF), and restriction fragment length polymorphism (RFLP) analysis with human or mouse, or with species-specific MHC-DNA probes. Species-specific probes were considered essential for the study of the genetic structure of the chicken and fish MHC. For the IEF technique allo-antisera, not necessarily species-specific, and monoclonal antibodies directed to monomorphic determinants of class I and II molecules are applied. Typing of many animals for association studies will most easily be resolved by serology. To understand and improve serology, IEF and RFLP both contribute

181

to, and, may be used as a tool for efficient production of reagents. Functional tests, like the graft-versus-host reaction and MLR, also provide a significant base for identification of MHC reagents. In poultry a new era of typing will start when oligonucleotide sequences corresponding to the allelic variants of the distinct MHC-molecules become available. In combination with the Polymerase Chain Reaction, oligotyping offers an interesting future for disease association studies. In order to speed up oligotyping prospects, Dr. Auffray proposed a cooperative effort to sequence one haplotype of the chicken MHC and to extend the analysis to the expressed genes for about 30 different haplotypes. The different techniques will remain important to understand the relationship between genes (expressed, pseudo), the gene products (class I and II molecules) and their serology (antigen binding sites), and function of the MHC.

In many laboratories two techniques at the most can be routinely handled. For our understanding of the structure of the MHC international cooperation utilising the local expertise is essential. Only then the number of loci and genes for each class of MHC-products for a particular species can be assessed and the linkage patterns resolved.

Biological function. Much more research should be devoted to understanding the immune mechanisms of the MHC in farm animals. Most relevant are studies on interactions between MHC molecules, processed and presented antigens and T-cell receptors. Successful application of new vaccines or vaccines based on recombinant DNA technology may be dependent on overcoming genetic restriction and bypassing the polymorphic nature of the MHC in outbred populations. The interactions with other genes (gene products) in or outside the MHC complex should also be included. Dr. Vaiman acknowledged in swine the presence of genes for complement factors Bf, C2 and C4 and the gene for the 21-hydroxylase enzyme, a main enzyme for steroid biosynthesis. The understanding of immune mechanisms in relation to disease, production traits, vaccination and stress is needed to identify the function of the MHC region and other "disease" genes in response to pathogens, and to provide a strategy for disease association studies. The technique of transfecting genes for MHC-molecules, T-cell receptors, or lymphokines has already lead to gene therapy in the mouse. This tool for the study of interactions is already being used in farm animals.

Experimental animal populations. There is a great need for pedigreed families and populations for the studies of the structure of the MHC (segregation analysis, population analysis) and for disease association studies. Gynogenetic fish, and inbred and congenic strains of poultry and swine have contributed greatly to the identification and characterization of the MHC in these species. For disease association studies in some species like fish and poultry, selection experiments with pathogenic challenges can show the involvement of the MHC. In other farm animals, for example in horse for sarcoid and goat for viral induced arthritis, family segregation analysis is the only feasible method, because no effects are seen at the population level. In general a more extensive MHC characterisation (class I and II, different loci) and correct clinical definition of disease is needed for further progress.

Parameters of disease. Mortality is a very global criterion for disease resistance. Knowledge of the pathogenesis of diseases should be directed towards improved clinical definition of disease. Improved clinical definition of specific diseases would also support the search for markers for disease by linkage studies.

Linkage maps. For identification of genes or gene markers for diseases and production traits, the present knowledge of linkage maps should be increased. In this respect fingerprinting with hypervariable DNA probes could be of great help, provided funds become available for this labour intensive work. Of course linkage maps only become useful when experimental populations of known susceptibility or resistance are available, and their crosses.

Loss of genetic variation. Breeding goals for disease resistance cannot be defined without definition of the future pathogenic environment. Selection for resistance to one disease favours certain MHC-haplotypes, as shown by Simonsen and coworkers, but how does the decrease in polymorphism affect the response to other, or newly arising pathogens? On the other hand, also shown by Simonsen, some heterozygous combinations are favoured, therefore supporting polymorphism in the population. Gene banks should be established to prevent loss of genes with possible interesting resistance functions. This need is even more urgent with the present strong emphasis on selection for production traits and maximal exclusion of natural selection. Because we do not know very much about genetic relations between production traits and diseases we cannot predict the loss of interesting genes, nor do we know which diseases will be most relevant in four or five generations.

Conclusions

European research on the MHC of farm animals and on genetics of disease has recently made much progress because of the application of new biotechnological techniques and the early investment in this area of research. For future success and application, international cooperation should be established to integrate diverse technical approaches, to exploit or organise experimental animal populations and to profit from increased understanding of the biological function of the MHC and pathogenesis of disease. International collaboration is a prerogative for the establishment of linkage maps or sequencing the genome, or parts thereof. Thus, there is a need for future meetings for exchange of ideas and data on the genetic basis of disease resistance in farm animals, and the loss of polymorphism of the MHC within species.

In conclusion: Research on animal disease genetics supports the advantageous use of genetic resistance and development of vaccines, and clarifies the relationship between productivity and disease. Furthermore this research profits from the enormous investment in medical research. New strategies for breeding for disease resistance have become available now, although they need to be continually and carefully considered in the light of future disease environments and of policies for vaccination in farm animals.

Participants

G.A.A. Albers   Euribrid B.V.
P.O. Box 30
5830 AA Boxmeer
The Netherlands

B. Amorena   Dept. of Genetics
Veterinary Faculty
University of Zaragoza
Miguel Servet 177
Zaragoza
Spain

E. Andresen   Dept. of Animal Genetics
The Royal Veterinary and Agricultural University
Bülowsvej 13
Copenhagen
DK-1870 Frederiksberg C
Denmark

G. Auffray   Institut d'Embryologie du CNRS et du Collège de France
49-bis Av. de la Belle Gabrielle
94130 Nogent sur Marne
France

H. Buschmann   Institut für Mikrobiologie
Infektionskrankheiten der Tiere
Veterinärstrasse 13
D-8 München 22
Germany

P. Chardon   Laboratoire de Radiobiologie Appliquée
INRA - CRJ
78350 Jouy-en-Josas
France

F. Coudert   Station de Pathologie aviaire et de parasitologie
INRA
Nouzilly
37380 Monnaie
France

I. Edfors-Lilja   Dept. of Animal Breeding and Genetics
Swedish University of Agricultural Sciences
P.O. Box 7023
S-75007 Uppsala
Sweden

E. Egberts   Dept. of Exp. Animal Morphology and Cell Biology
Agricultural University
Marijkeweg 40
6709 PG Wageningen
The Netherlands

W. Hartmann
Institut für Kleintierzucht
BFAL Braunschweig-Völkenrode
Dörnbergstraße 25/27
D-3100 Celle
Germany

E. Hensen
Dept. of Immunology
Faculty of Veterinary Medicine
University of Utrecht
P.O. Box 165
3508 TD Utrecht
The Netherlands

P. Kaastrup
Institute for Experimental Immunology
University of Copenhagen
Nørre Allé 71
DK-2100 Copenhagen
Denmark

G. Koptopoulos
Aristostelian University
Faculty of Veterinary Medicine
Department of Microbiology and
Infectious Diseases
54006 Thassaloniki
Greece

B. Kristensen
Dept. of Animal Genetics
The Royal Veterinary and Agricultural University
Bülowsvej 13
DK-1870 Copenhagen, Frederiksberg C
Denmark

F. Lantier
Station de Pathologie de la Reproduction
37380 Nouzilly
France

S. Lazary
Institut für Tierzucht
Abteilung für Immunogenetik
Bremgartenstraße 109a
CH-3012 Bern
Switzerland

W. Leibold
Immunologie der Tierärztlichen Hochschule Hannover
Bischofsholer Damm 15
3000 Hannover 1
Germany

P. Madsen
National Institute of Animal Sciences
P.O. Box 39
DK-8833 Ørum Sønderlyng
Denmark

D. Meggiolaro   Istituto di Zootecnica Generale
                     Facolta di Agraria
                     Via Celoria 2
                     20133 Milano
                     Italy

T. Naglic   Veterinary Faculty
                     Greyzilove 55
                     41000 Zagreb
                     Yugoslavia

K. O'Farrell   Moorepark Research Centre
                     Fermoy
                     County Cork
                     Ireland

H. Østergard   Dept. of Animal Genetics
                     The Royal Veterinary and Agricultural University
                     Bülowsvej 13
                     DK-1870 Copenhagen, Frederiksberg C
                     Denmark

J.J. van der Poel   Dept. of Animal Breeding
                     Agricultural University
                     P.O. Box 338
                     6700 AH Wageningen
                     The Netherlands

M. Simonsen   Institute for Experimental Immunology
                     University of Copenhagen
                     Nørre Allé 71
                     DK-2100 Copenhagen
                     Denmark

R.L. Spooner   AFRC Physiology and Genetics, Research Station
                     King's Buildings
                     West Mains Road
                     Edinburgh EH9 3JQ
                     Scotland

W. Sybesma   IVO "Schoonoord"
                     P.O. Box 501
                     3700 AM Zeist
                     The Netherlands

M. Vaiman   Laboratoire de Radiobiologie Appliquée
                     IPSN-DPS-SPE
                     9191 Sif sur Yvette
                     Cedex
                     France

H. Varewyck   Rijksuniversiteit Gent
Leerstoel Dierlijke Genetica en Veeteelt
Heidestraat 19
9220 Merelbeke
Belgium

E. Viñuela   Centro de Biología Molecular
Consejo superior de investigaciones científicas
Madrid
Spain

R.R.P. de Vries   Afd. Immunohaematologie en Bloedbank
Academisch Ziekenhuis Leiden
Gebouw 1, E3-Q
Rijnsburgerweg 10
2333 AA Leiden
The Netherlands

J. Williams   AFRC Physiology and Genetics, Research Station
King's Buildings
West Mains Road
Edinburgh EH9 3JQ
Scotland

M. Zanotti Casati   Istituto di Zootecnica
Facolta di Medicina Veterinaria
Universita degli Studi di Milano
Via Celoria 10
Milano
Italy

A.J. van der Zijpp   Dept. of Animal Husbandry
Agricultural University
P.O. Box 338
6700 AH Wageningen
The Netherlans

Current Topics in Veterinary Medicine and Animal Science

Recent publications

1984

26. Manipulation of Growth in Farm Animals, edited by J.F. Roche and D. O'Callaghan. ISBN 0–89838–617–8
27. Latent Herpes Virus Infections in Veterinary Medicine, edited by G. Wittmann, R.M. Gaskell and H.-J. Rziha. ISBN 0–89838–622–5
28. Grassland Beef Production, edited by W. Holmes. ISBN 0–89838–650–0
29. Recent Advances in Virus Diagnosis, edited by M.S. McNulty and J.B. McFerran. ISBN 0–89838–674–8
30. The Male in Farm Animal Reproduction, edited by M. Courot. ISBN 0–89838–682–9

1985

31. Endocrine Causes of Seasonal and Lactational Anestrus in Farm Animals, edited by F. Ellendorff and F. Elsaesser. ISBN 0–89838–738–8
32. Brucella Melitensis, edited by J.M. Verger and M. Plommet. ISBN 0–89838–742–6

1986

33. Diagnosis of Mycotoxicoses, edited by J.L. Richard and J.R. Thurston. ISBN 0–89838–751–5
34. Embryonic Mortality in Farm Animals, edited by J.M. Sreenan and M.G. Diskin. ISBN 0–89838–772–8
35. Social Space for Domestic Animals, edited by R. Zayan. ISBN 0–89838–773–6
36. The Present State of Leptospirosis Diagnosis and Control, edited by W.A. Ellis and T.W.A. Little. ISBN 0–89838–777–9
37. Acute Virus Infections of Poultry, edited by J.B. McFerran and M.S. McNulty. ISBN 0–89838–809–0

1987

38. Evaluation and Control of Meat Quality in Pigs, edited by P.V. Tarrant, G. Eikelenboom and G. Monin. ISBN 0–89838–854–6
39. Follicular Growth and Ovulation Rate in Farm Animals, edited by J.F. Roche and D. O'Callaghan. ISBN 0–89838–855–4
40. Cattle Housing Systems, Lameness and Behaviour, edited by H.K. Wierenga and D.J. Peterse. ISBN 0–89838–862–7
41. Physiological and Pharmacological Aspects of the Reticulo-rumen, edited by L.A.A. Ooms, A.D. Degryse and A.S.J.P.A.M. van Miert. ISBN 0–89838–878–3
42. Biology of Stress in Farm Animals: An Integrative Approach, edited by P.R. Wiepkema and P.W.M. van Adrichem. ISBN 0–89838–895–3
43. Helminth Zoonoses, edited by S. Geerts, V. Kumar and J. Brandt. ISBN 0–89838–896–1
44. Energy Metabolism in Farm Animals: Effects of Housing, Stress and Disease, edited by M.W.A. Verstegen and A.M. Henken. ISBN 0–89838–974–7
45. Summer Mastitis, edited by G. Thomas, H.J. Over, U. Vecht and P. Nansen. ISBN 0–89838–982–8

1988

46. Modelling of Livestock Production Systems, edited by S. Korver and J.A.M. van Arendonk. ISBN 0–89838–373–0
47. Increasing Small Ruminant Productivity in Semi-arid Areas, edited by E.F. Thomson and F.S. Thomson. ISBN 0–89838–386–2
48. The Management and Health of Farmed Deer, edited by H.W. Reid. ISBN 0–89838–408–7

1989

49. Vaccination and Control of Aujeszky's Disease, edited by J.T. van Oirschot. ISBN 0–7923–0184–6
50. Pathological Histology of Domestic Animals, edited by J. von Sandersleben, K. Dämmrich and E. Dahme. ISBN 0–7923–0311–3
51. Diagnostic Ultrasound and Animal Reproduction, edited by M.A.M. Taverne and A.H. Willemse. ISBN 0–7923–0403–9
52. Improving Genetic Disease Resistance on Farm Animals, edited by A.J. van der Zijpp and W. Sybesma. ISBN 0–7923–0518–3